A SCIENTIST'S TOOLS FOR BUSINESS
Metaphors and Modes of Thought

A SCIENTIST'S TOOLS FOR BUSINESS
Metaphors and Modes of Thought

ROBERT L. SPROULL

UNIVERSITY OF ROCHESTER PRESS

First published 1997

University of Rochester Press
668 Mt. Hope Avenue
Rochester, NY 14620   USA

and at P.O. Box 9
Woodbridge, Suffolk IP12 3DF
United Kingdom

ISBN 1–878822–84–5

Library of Congress Cataloging-in-Publication Data
Sproull, Robert L.
    A scientist's tools for business : metaphors and modes of thought / Robert L. Sproull.
        p.   cm.
    Includes index.
    ISBN 1-878822-84-5 (alk. paper)
    1. Management.   2. Management science.   3. Science.
I. Title.
HD31.S63   1997
658—dc21   97-30781
                                            CIP

British Library Cataloguing-in-Publication Data
A catalogue record for this book is
available from the British Library

Designed and typeset by Cornerstone Composition Services
Printed in the United States of America
This publication is printed on acid-free paper

Contents

1

Introduction

The purpose of this little book is to illustrate and explain the *connections* between business problems and the tools and modes of thinking of the scientist. I hope to make accessible to nonscientists some of these approaches that the scientist has found to be indispensable.

The primary readers whom I address are any thoughtful persons without scientific experience and perhaps with only unpleasant memories of science and mathematics in school, but I specifically aim to help the corporate executive and aspiring executive.

Another group who should find these connections useful are people who have been educated as engineers and who are becoming managers. They may not have realized the variety and strength of applications to business problems of the tools they were familiar with in their backgrounds.

Of course, my hope is that a broader audience—in fact anyone who interacts with people and operations in some administrative or cooperative role—will find here a useful extension of his or her thinking. Since most executives and others in business also interact with the not-for-profit world, some of my examples illuminate management in that world, also.

Some of these tools are the scientist's ways of using mathematical reasoning without becoming bogged down in detailed mathematics. Mathematics, "the handmaiden of the sciences," can also be a powerful servant to the executive. The mathematics presented here is entirely oriented toward applications, with none of the logical derivations and elegance that often discourage nonscientists from applying even elementary mathematics. Other tools are scientific metaphors that are applicable to an executive's analysis and decision making. Perhaps I should defend here my appropriation of the term "metaphor."

Literature abounds with examples, illustrations, and metaphors that enrich our experience, enlarge our understanding, and deepen our thinking. They also serve as communication vehicles and constitute part of the difference between education and mere training.

> I can call spirits from the vasty deep.
> Why, so can I, or so can any man;
> But will they come when you do call for them?
> (*King Henry IV*, Part I, Act 3)

One need not have read Shakespeare to have achieved a healthy skepticism about the claims of prophets, promoters, and some politicians, but the shared memory of that quotation is helpful in our thinking and communicating.

And it is not only thinking and communicating that are aided and enriched by literature. For several years I was superbly served by a secretary who was extremely competent but who hated people. What kept me from unproductive depression was remembering King Gama's song in *Princess Ida*:

> O, don't the days seem lank and long,
> When all goes right and nothing goes wrong,
> And isn't your life extremely flat,
> With nothing whatever to grumble at!

I do not claim that science and engineering provide as rich an array as literature. But I do claim that there are many metaphors and ways of thinking and analysis used by scientists and engineers that could help the nonscientist. (After this Introduction, to avoid repeating "scientists and engineers" I shall say only "scientist," but the engineer's contribution is at least equal.) These are largely hidden in textbooks and technical articles, although some are so simple and universal that they are just assumed and not written down. The simplest, not peddled here, are merely communication vehicles, like the "litmus test" and "tip of the iceberg" commonly found in newspapers. The aids we shall explore here go well beyond communication and enlarge our thinking.

These tools are not in any way better than the rules, procedures, and metaphors emanating from literature and philosophy. But you do not have to choose: You can add any of these you like to your tool bag without taking anything out.

I apologize in advance for any instances of underestimating the reader and "teaching my grandmother to blow eggs." I am sure almost any reader will find some of this material familiar, but I hope there will be enough new thoughts and approaches to justify the effort of reading.

I hope the reader will not object to the leavening of this solid stuff by the occasional introduction of anecdotes. These are often only peripherally related to the material at hand, but they should carry some weight (I can hardly claim wisdom) of their own.

In addition to tools and metaphors, I have thought it wise to include some cautions and warnings, particularly in areas of common usage like statistics. Armed with these, the reader will be better prepared to make his or her own choice from the cafeteria of "scientific" offerings from his associates or from the press. These and some other topics should also help the executive in managing the engineering and engineers involved in product and process development.

The origin of this book comes from many years of moving back and forth among teaching, science, engineering, corporate management, and governmental and educational administration.

In interacting with many very capable but non-technical people, I gradually learned that there were useful patterns of thinking, approaches to problems, and communication tools that the scientist and engineer assumed everyone used but which were in fact used only in our own, parochial sphere.

I should also explain what this book is *not*. It is not a glossary of jargon and buzzwords. It does not dish out the cute language of "computerspeak." It does not provide any of the *substance* of science and engineering (well, maybe just a tiny bit); it does not even speak to "the scientific method." It does not attempt to provide cocktail party conversation or defenses against the "lifemanship" exhibited at such parties that is based on a superficial acquaintance with trendy subjects like black holes or the taxonomy of the particles of high-energy physics. If the book *impresses*, I have failed; my intent is to *inform*.

Acknowledgments

My principal and very important debt is to scores of colleagues whose problems or questions elicited these thoughts. Most of these were corporate executives and directors in the years when I served on corporate boards. Many were also colleagues in universities or the federal government.

I also appreciate the comments and suggestions by several readers of preliminary forms of the manuscript, notably Andrew H. Neilly, Jr., Erik M.Pell, Robert F. Sproull, and especially Robert S. MacDonald.

2

How to Specify Just Enough

Let's start with an easy one. This chapter starts with *communication*, but thinking a little about the communication process quickly leads us to consideration of what to *communicate* and what to *measure*.

Consider a corporate financial statement. A small corporation may report a profit of, say, $7,275,536; a large corporation may report a "profit in thousands" of $235,407. Is there anything wrong in including all those digits in those big numbers?

No, if you are doing accounting for the prevention of cheating and fraud, for taxation, or for paying bills. All of us expect our banks to give us statements accurate to the cent.

Yes, emphatically, it is certainly wrong if the financial report is intended to communicate to shareholders, employees, and customers the performance of the corporation. Reporting $7,275,536 suggests that we know that the profit was not, say, $7,275,528; it claims a totally misleading *precision* of measurement. The scientist would say that because of many complexities and imponderables in arriving at the profit, most of those digits are *insignificant*. After consideration of these complications, we might conclude, for example, that all we know is that the profit was between about $7,200,000 and about $7,400,000; the scientist would report $7,300,000.

Even cash flow, a concept much more precisely defined than profit, can benefit from some rounding. In large companies, with multiple divisions, it may be necessary to keep more detail in the figures for interdivisional transfers.

Does not this reporting of only *significant* figures hold back some information by denying access to those dropped digits? Not at all. $7,300,000 actually provides *more* information than $7,275,536. It not only tells the profit to as many figures as it is worth paying attention to, but it also tells us what *precision* the managers claim for that number. If, for example, there were major uncertainties in new product acceptance, warranty expense, or litigation, the profit might have been properly reported as $7,000,000. Of course, it may be some time before the Securities and Exchange Commission will rise to this level of sophistication, and SEC and other regulations may prevent a proper job of external reporting. But as managers inside an enterprise, we should not fool ourselves.

It is this extra step of estimating the meaningful precision that makes reporting only the *significant figures* most powerful. The managers must pause and ask themselves: Do I know this to 0.001% or only to, say, 1%? Avoiding misleading the public is worthwhile enough, but forcing thought on the meaning of the figures for management is even more valuable.

The concept and the practice of significant figures play an even more serious role in corporate *planning*. I have seen five-year plans in which the numbers were reported (in the fifth year!) as if they were known to one part in ten million. Interestingly enough, in those same plans the key assumptions about the inflation rate, currency conversions (e.g., yen to dollars), taxation rates, consumer spending, and the like were not reported at all! (The assumptions about all such quantities should clearly be explicit in plans, in order that the "actual" may be properly compared with the plan, as time goes on and the plans are appropriately modified.) Although I should like to call "Foul!" for misleading the reader, I limit my complaint to alleging that planning was inadequate, in that no estimate of precision or the range of outcomes was provided.

Not only does the restriction to significant figures avoid misleading others and necessitate estimating meaningful precision, but it results in a *compact* reporting. You do not even require the space to write $7,300,000±100,000 and yet you convey that information.

Incidentally, although in this example the scientist's love for compactness can produce a benign result, that love is a major reason the general public finds science arcane and why the media writers are "turned off" by scientific articles. A scientist may grope and stumble and stagger up many blind alleys, but when he is satisfied with his work and writes it into the scientific literature, he records the surviving part in orderly, extremely compact (and therefore hard-to-read) form. Too bad.

Misleading by *insignificant figures* is not confined to corporate reporting; the newspapers are full of it. For example, a terrorist bombs a restaurant in Germany. A reporter asks a witness: "How big do you think the bomb was?" Answer: "I don't know—it's such a mess in there—but it must have been something like a kilogram." The American newspaper, after the reporter looked up "kilogram" in his dictionary, publishes: "The bomb was reported to be 2.2 pounds." The reader suffers from a wholly false idea; he concludes that the investigators knew or learned a great deal in detail about the bombing to report the bomb's size within 5%. This variety of nonsense is widespread and occurs especially frequently with currency translations.

Precision and significant figures expose the question of the purpose of accounting. Much agony and confusion arise from failure to distinguish accounting for the prevention of cheating and fraud (and keeping the Internal Revenue Service and the SEC happy) from accounting for the purpose of managing; the ubiquity of and necessity for the former often distracts us from the needs of the latter. Accounting for tax and banking purposes must be to the dollar and is circumscribed by regulations. Accounting for *managing* must be timely and to only a meaningful precision, with sensible estimates of that precision.

The usual corporate reporting sadly mixes the two kinds. Anything that can be measured in dollars is reported to a monstrous number of digits. Worse, no reporting at all is provided for those characteristics, most of which can be estimated quantitatively but cannot be expressed in dollars, that are more important for understanding the vitality and promise of the corporation: success in recruiting and retention of staff; age of machine tools; the corporation's track record of estimating market size and market share; the quality and relevance of research and development; and many more. Some of these (e.g., market share by product line) are necessarily confidential, but failure to address any of them while indulging in outlandish digitry in dollar figures diverts attention from the true evaluation of the corporation.

In annual reports, important considerations like market share and development successes are usually given an optimistic but vague treatment in the chairman's text and then some very forbidding cautionary remarks in fine print in the "financials." There is a great middle ground between these extremes. Although it may not be wise to fill it all in the annual report, it is vital that management give it serious analysis.

Incidentally, scientists early in their careers usually fall into the same trap of failing to employ appropriate accounting but in a quite different way. Most university scientists have support from grants, usually from the federal government but often from states or companies. The university accountants are the formal administrators of the grants and give each scientist monthly reports, accurate to the penny but usually late and never including, of course, the scientist's own plans for supporting students, buying equipment, and incurring other obligations. The young scientist begins by complaining bitterly that central administration, far from helping him, is actually hindering his management by giving him misleading reports. The successful scientists soon learn that for their own management purposes they must do their own, "back of the envelope" accounting, which may be accurate to only a few percent but which is up-to-date and includes their planned expenditures as well as two-month-old bills.

Your development engineers play out the same scenario with the budget you give them.

I introduced the message of significant figures, using appropriate precision to specify just enough, through the question of corporate reporting of dollar figures and then *generalized* by bringing in the other measures of corporate health. The message can be further generalized by application to *any* measurement question in manufacturing, marketing, product performance, or even personnel performance. You can always profitably ask: What is the worthwhile, meaningful precision of each quantity? Exposing this question early saves useless cost and time and, more importantly, stimulates consideration of what are the most significant quantities.

To generalize in a different direction, another application of appropriate precision is to the time interval at which observations are made or conclusions are reached. Again, attempting to be too precise about time can lead to bad management. For example, in visiting your research and development laboratories it is a mistake to visit too often: If you visit once a week you learn only whether some equipment has arrived. If you visit once a year you can learn about selection of areas of concentration, strategy, risks, comparison with competitors, and other important features.

Scientists commonly use this same move from the specific to the general. There is, of course, no logical proof that what happens (or is worthwhile) in a specific case will happen (or be worthwhile) generally, over a broader front. But understanding a specific, often simple, case is *suggestive* of a generalization, which then must be proved or disproved on its merits. This path is called *heuristic* reasoning, contrasted against logically compelling reasoning. Science is described and taught as if it always proceeded from general laws to specific cases, but the *development* of science usually was the other way around. We shall explore an especially interesting example of this mode of thought in Chapter 16.

Meanwhile, there are opportunities for you to take advantage of this kind of reasoning in your business. Frequently a special case, a particularly—perhaps accidentally—successful product or product promotion can suggest a generalization (similar treatment) to a whole product line or to quite different products.

Not all numbers are susceptible to reduction to significant figures. Only numbers that report measurements are so susceptible; some numbers report addresses, and they cannot be other than literally copied. I once knew a secretary who thought acceptable performance was getting a telephone number correct to 1%!

This chapter has explained the scientist's way of looking at the world, always with a ± in mind, always making sure that *all* the relevant measurements have been included. You can profitably follow that pattern in assessing the health of your company or the success of your program.

3

Change and Growth

Salome [Arizona] always has kept up its Average Annual Growth of 100% a Year—19 People now in 19 Years—but after going through the Panic of 1907, the World War and 3 Democratic Asphyxiations without a Slump, it looks as if this here Greasewood Golf Course was going to depopulate the town."
[Dick Wick Hall in *The Best of the West*, edited by Tony Hillerman.]

Now what can you make of all those percents? Clearly not much, and you are not intended to in this spoof about a twenty-mile desert golf course. But seriously intended use of percentage often is equally confusing.

Percentage is commonly used to report change and growth, which are among the quantities that an executive most often needs to measure and to report, as in: "the Consumer Price Index rose 3.2% last year." Now, language, plain English as well as scientific language, provides powerful alternatives to percentage; nevertheless we are imprisoned by the widespread, often misleading, use of percent. Measuring and reporting change and growth are of great importance to the executive, and so in this chapter we shall explore some of the alternatives to an uncritical use of percentage.

I should acknowledge, as we begin, that percent has its proper place and serves us well if it is confined to that place. A bond yield of 6.2% or an annual inflation rate of 3.5% is easy to understand and unambiguous.

But percentage becomes treacherous when the numbers grow, and especially when percentage is used to express both increases and decreases.

First, the large numbers: suppose something that had been $100 billion (e.g., U.S. health costs over a particular decade) was said to have "increased by 200%." Is it now $200 billion or $300 billion? The reader cannot tell since both expressions are used, even occasionally by the same writer. Confusion reigns.

And what is one to make of a statement that some activity has been "slowed 100%"? This language is not usually intended to imply a stop but only a reduction to one-half of the original value.

Next the up-and-down problem: If at some time the Canadian dollar equals $0.70 U.S., it is equally accurate to say that the Canadian dollar sells at a discount of 30% from the U.S. dollar and that the U.S. dollar sells at a premium of 43% over the Canadian dollar. If the language is carefully chosen, each can be used. But confusion is likely, especially if fueled by avarice or ignorance.

At a time when $1 Canadian equaled $0.70 U.S., a hotel clerk in rural Canada changed my U.S. dollars by first (appropriately) deducting his transaction fee and then (inappropriately) giving me $1.30 Canadian for each U.S. dollar. When I protested, he easily procured my agreement that "the exchange was 30%," but I contended that the percentage he quoted was intended to apply in the other direction. He refused to undertake the "thought experiment" of converting U.S. dollars to Canadian and then back again to U.S. by his method; I think he suspected a trap or that I was a confidence man. I had to give up; it was only money, and the clincher was, "Do you want the room or don't you?"

In the Great Depression of the 1930s, a major university cut faculty salaries 10%. Several years later, the university administration announced that it had "restored" the salaries by raising salaries 10%, returning, of course, to only 0.99 times the original salaries. Hell hath no fury like a university faculty suspecting its central administration of fraud, even if the difference (as in this case) was only 1%!

This up-and-down confusion has had serious consequences for many small building contractors. Across the industry, the typical *margin* added to the cost of labor and materials is 33% of the contract price, to cover management, bidding, and other indirect costs. But that means that a *markup* of 50% must be added to the labor-and-materials invoices. The scene is littered with the failed companies that added 33%.

On a less serious note, there is an old story about the extremely successful alumnus who was speaking at the dedication of a mathematics building he had given to his alma mater. Rather modestly deprecating his business achievement, he said: "All I did was make those widgets for $1.00 and sell them for $3.00, but those three percents sure did mount up!"

What are the alternatives when the numbers become large or when one wishes to go both ways accurately? The best one for general use is to speak only of *factors*, as, for example: "The health costs over that decade increased by a factor of 3." Note that if they ever decreased (unlikely!) from $300 billion to $100 billion, one would say they "decreased by a factor of 3." By this language one avoids the confusion of calling it a decrease of 67%, or 300%, or of whatever other misuse to which percentage can be put.

The "factor of" approach can be used for *any* increase or decrease without confusion, and its general use would be very healthy. Obviously one could still use percent appropriately for *small* changes; although a "factor of 1.05" would be accurate, "a 5% increase" would be equally accurate and would serve as well.

I shall next describe the scientist's way of dealing with increases and decreases by large factors since it will be useful for our later work. I do not claim that it should enter general use, and you may wish to skip the remainder of this chapter.

This usage developed naturally from dealing with very long telephone lines, on which electrical signals are attenuated by large factors and amplifiers are inserted periodically to bring the signals back up to their original strengths. Both attenuation and amplification are expressed as so many *db*, pronounced "deebee" and originally called "decibels." If the signal's power decreased by a factor of 10, say, from 10 watts to 1 watt, it is said to have "lost 10 db" or "decreased by 10 db." If it has decreased by a factor of 100, say from 10 watts to 0.1 watt, then it is said to have "lost 20 db." A factor of 1000 is 30 db, and so on. It is convenient to remember that a factor of 2 is about 3 db.

If you simply wish to memorize the fact that 10 db is a factor of 10, you will know all that you need to know to understand the rest of this book and you can safely skip the next four paragraphs. (Both the preceding paragraph and the next four paragraphs will be made clearer when we discuss graphs in Chapter 5.)

You will probably have noticed that the number of db is simply 10 times the logarithm to the base 10 (the "common logarithm") of the factor; for example, the $\log_{10}$ of 100 is 2, and 10 times 2 gives us 20 db. You will almost certainly have been subjected to attempts to teach about logarithms, usually in high school. Logarithms were traditionally taught as a means of multiplying or dividing large numbers with high precision. You will almost certainly have completely forgotten about them because no one (well, almost no one) multiplies or divides to a high precision and also because any such activity would now be done with a $5 handheld calculator, not by the laborious use of logarithms.

But the *concept* of a logarithm is too useful to abandon. The logarithm to the base 10 (call it L) of a number N is defined by $10^L = N$ and is written $\log_{10}N$; or in words, the logarithm of a number is the power we have to raise 10 to in order to get that number. The magic of logarithms derives from the fact that to multiply two numbers when they are expressed as exponents of the same number, one can merely add exponents. Try it: $2^2 = 4$; $2^3 = 8$; $2^2 \times 2^3 = 4 \times 8 = 32 = 2^{2+3}$. Or, more to the point for our application here, $10^2 = 100$ and thus $\log_{10}100 = 2$; $10^3 = 1000$ and thus $\log_{10}1000 = 3$; $10^2 \times 10^3 = 100 \times 1000 = 100,000 = 10^{2+3} = 10^5$. Adding the logarithms gives the logarithm $(2 + 3 = 5)$ of the product; the product itself is then recovered by raising 10 to the power 5.

If it is not immediately apparent how this works, rereading is not likely to help. I suggest playing with it a little while *with a pencil and paper*, trying various examples. If you make a few multiplications where you know the answer, the power of logarithms should become apparent.

You do not have to think about logarithms to find db useful, but the logarithmic nature of db is responsible for their valuable property. As factors multiply one another, the db representing these factors *add*. For example, if a telephone line has an attenuation of 3 db per mile, the signal is 1/2 of its original value after 1 mile, 1/4 after 2 miles, 1/8 after 3, and so on. We say that the signal "loses 3 db" in each mile; it loses 30 db in 10 miles and 60 db in 20 miles (i.e., it is decreased by a factor of 1,000,000). It then requires an amplifier with a gain of 60 db (again a factor of 1,000,000) to restore the signal to its original strength.

You have probably encountered db in connection with noise. Since the sound pollution we are necessarily subjected to covers many factors of 10 ("orders of magnitude"), it is hopeless to talk about it in terms of percent (we should have to say that a power lawnmower increases the noise from a background of wind in the trees and bird songs by figures like 1,000,000%!). The accepted practice is to set an arbitrary noise standard and to express each environment as so many db noisier than that. An artificially produced noise that approximates the threshold of hearing of an "average" person is agreed upon as the standard for 0 db; the scale is then accurate and precise for differences in sound level as measured by physical instruments. Fifty feet from a power lawnmower then would come in at about 70 db; 1000 feet from a jet aircraft taking off, at 90 db; and a front table at a discotheque, at an ear-shattering 120 db.

It would be convenient and less error prone if all large changes (e.g., the Dow-Jones Industrial Average over a secular period like thirty years) were expressed in db. The symmetry of up and down movements and using the addition of numbers of db to produce products of factors could be very helpful. It would have the incidental advantage, if you recall the preceding chapter, that one would not see absurdities like the television reporting of a change such as 4.67 in the DJIA when it was in the neighborhood of 7000, however useful that might be for floor traders. (For no reason that I can think of, the situation is even more ridiculous with the Wilshire 5000, reported every evening to an absolutely mythological precision like 7275.657.) But I am sure that stock market averages will *not* be reported in db; that practice just sounds "too mathematical" for general consumption.

My pessimism on this score is enhanced by my experience of failing to make a university dean's budget cut less objectionable by telling him it was "only 0.2 db." It was the dean of the College of Engineering, and he could tell a 5% cut, however dressed up in scientific language. I got nowhere.

Before leaving percentage, I should like to mention one very helpful use of percentage when the same, small percentage change occurs cycle after cycle (usually an annual percentage change, year after year). This is the Rule of 72 to estimate when a change by a factor of 2 has occurred. The rule is: Divide 72 by the percentage change per cycle.

If, for example, a security yields 6% per year, in how many years will it double in value if all the 6%'s are applied to increasing the principal? The answer is $72 \div 6 = 12$ years.

Strictly speaking, "72" should be "69," and this rule works to a percent or so of accuracy only for small (less than 10%) annual increments. But $72 = 3^2 \times 2^3$ is easier to work with (making calculations in your head) since you can frequently cancel some of its many factors. Using 72 instead of 69 happens to be a little *more* accurate for the range 5% to 10%, whereas 69 is accurate for much smaller percentages. (Much later, in Chapter 13, we shall see where this number 69 comes from.)

What can you carry from this chapter? First, use of percentage to describe both increases and decreases or to describe large changes is at best misleading and at worst just plain wrong. Second, the specification of changes by a "factor of" avoids these pitfalls. Third (if you have survived that section), logarithms can be friendly, and the use of db based on logarithms can be very helpful in dealing with quantities (like environmental noise) that vary over many orders of magnitude. Finally, the Rule of 72 is a very handy way of computing the number of cycles (number of years if one has a yearly increment) required to produce a change of a factor of 2. If you did not study the preceding ten paragraphs, all you need for the rest of this book is that a change of 10 db means a factor of 10; of 20 db, a factor of 100; of 30 db, a factor of 1000; and so on.

4

The New Mechanics

Some of the most important questions facing an executive are the choice of the units or aggregates to pay attention to and what to measure in order to have a sound basis for management. Tackling either question can benefit from a quick look at the new mechanics of the twentieth century. In addition, the new mechanics provides some metaphorical insight into the questioning process.

The new mechanics has unfortunately acquired a reputation of being very difficult to understand and to manipulate. To be sure, in its microscopic detail it does require deep preparation and thought, and there are still marvelously intricate unresolved problems in its application to the submicroscopic world. But the implications for our real, touchable world are simple and yet powerful.

Physics until about 1900 was accustomed to a world in which *continuity* was featured almost exclusively. If, for example, the quantity in question was motion, then the position, velocity, acceleration, and other attributes were always continuous, that is, there were no abrupt jumps, no failures to exist at one point and abrupt reappearance at another. All of our experience and our observations with our own eyes support the view of a continuous physical world.

Thus it could have been expected that a large amount of intellectual trauma would ensue when overwhelming evidence accrued from atomic spectra (light emitted by excited atoms) and other experiments which demonstrated conclusively that on the atomic scale of sizes and energies the world was basically *discontinuous*. It was discovered that an atom can exist in only one or another of a set of states with *discrete* energy levels, like a staircase but with unequal heights of the treads. When an atom goes from a higher to a lower such level (called a *quantum state*), it emits a packet of energy called an *energy quantum*. We say that energy is *quantized*. The emitted light appears in bunches called light quanta or *photons*.

All of these features and many more were integrated in the 1920s into a beautiful theory called *quantum mechanics*. This theory *explained* the observed phenomena of physics and chemistry, but far more importantly, it led onward with understanding and *predictions* for a vast array of new phenomena. (A scientific theory that only explains, without predictions that can be compared with observations, is untested and devoid of practical utility.) Among a host of accomplishments, I cite only one group: The whole collection of solid-state devices that make modern computing and communication possible originated from investigations stimulated and guided by quantum mechanics.

I have raised the subject of quantum mechanics in order to draw out three applications, of which the second and third are closely related.

The first has already been introduced, the concept of quantization into discrete packages. You are already familiar with this concept in your examination of capital projects like buildings or new production facilities. You know that you must deal with each of these as a substantial quantum. Any attempt to "get a foot in the door" with first an approval of the land, then the building, then the tools and equipment, and then the training and transfer expenses has to be successfully resisted.

Although an old production facility can be more or less continuously renovated and upgraded, there can be substantial advantages in quantizing a modernization program into indivisible major units, each addressed to a specific goal and each comprising *all* the costs, from acquiring land to moving-in expense. Each quantum can then get the scrutiny it deserves, with an examination of probable payback periods and a study of alternatives.

Quantization appears in its metaphoric sense in many and sometimes strange places. My favorite example is the tremendous down comforter that makes sleep possible in a cold Bavarian bedroom. It is a delightful device with the insulating power of three or four blankets but only the stability of an upside-down waterbed when floating over my recumbent body. But this covering comes quantized: I either have it or I do not. Unlike three blankets, any one of which I could remove if I am too warm, I must remove the entire comforter or leave it.

You will profitably pay considerable attention to how you can best quantize the activities that make up your business. The aggregation of pieces into manageable units of appropriate size and limited complexity is necessary in all but the smallest enterprises. One consideration is always how many others can report to an executive, and although a digression from quantum mechanics, there is an interesting metaphor from the chemistry and physics of crystals. The number of nearest neighbors of a particular atom in a crystal is called the "coordination number." This language leads to speaking of the "coordination number of the throne," the number of other executives who report to the central one.

The second application of quantum mechanics involves the famous *Uncertainty Principle*. I raise this primarily to produce a metaphor about measurement but also in part to caution you against the popular view of alleged "uncertainty." This principle should actually be called the *Indeterminacy Principle* since in its original German it was the *Unbestimmtheitprinzip*. You may well wonder why I make a point of using an uncommon word, "indeterminacy," in place of the familiar "uncertainty." My reason is that there is *nothing* uncertain about the predictions and explanations of quantum mechanics when applied to real situations; using the common word tempts us to a metaphorical analogy between some uncertain situation in our immediate experience and a physical principle. The word "indeterminacy" is so uncommon and infelicitous that we are not so tempted.

There is much more at stake here than just hairsplitting over words: There is a burgeoning literature attacking solid scientific facts and scientific contributions to understanding and human welfare that uses the alleged uncertainty of quantum mechanics predictions in real situations as if it vitiates the success of science.

The Indeterminacy Principle states that there are pairs of variables (like energy and time) such that both cannot be simultaneously measured with unlimited accuracy. If I wish to measure or specify one accurately I can measure or specify the other only to within a range (accurately predicted!) of inaccuracy. If I wish to be clear and accurate about the energy level of an individual atom, I cannot specify exactly when it makes a transition from one state to another. If I wish to be clear and accurate about the transition time, I must accept a range of possible energy states. The process of measuring each of a pair of variables (such as energy and time in the atomic example, or the position and momentum of an electron) alters the environment for measuring the other. This interaction between the measuring process and the quantity being measured is a feature of quantum physics that is quite different from the earlier ("classical") physics.

I learned early in my experience as an administrator that if I were obviously taking down notes in a one-on-one meeting, the other person spoke more carefully and responsibly. My "measurement" evidently altered the quantity being measured.

You will know of many instances where the presence of an observer at a meeting altered what would have happened in his or her absence. The arrival of a television camera or representatives of the press dramatically changes an informal give-and-take session into a formal affair, perhaps featuring much posturing and even shouting.

Perhaps the most famous examples of altering the situation to be measured or observed by the act of measuring or observing are the "Potemkin villages." Catherine the Great's minister Potemkin constructed whole model villages, or at least the part of them facing the high road, so that on the Empress's "fact-finding" trip through the provinces she would observe only happy, prosperous villagers. The repetition of such an egregious case is now unlikely since the advance men of the TV networks would surely discover the fraud.

But there is a related danger in the modern corporation where the headquarters are physically separated from all the operations. The executive at international headquarters in Manhattan or southern Connecticut quite properly says that he or she can be on the company's jet early in the morning and spend most of the day at any of the company's installations. But all such trips are *planned*, and the executive's presence is predicted. The observations and interactions are useful and necessary, but they are not the same as if corporate headquarters were at one of the company's plants and the CEO, upon running into a colleague from the plant in the corridor, could go off to lunch with him or her.

This second application highlights the way choosing to measure some aspect of what you are managing alters the activity. Even asking questions has consequences. It is a well-known adage in business that the troops pay attention to what is measured and reported to the top. I have already (in Chapter 2) alluded to the measurements that are missing from annual reports. If you measure success in recruiting and retaining staff as quantitatively as possible, you will alter the personnel (human resource) process. If you measure and have reported to you the number of patents obtained, you will alter the priorities of your development engineers.

The third application is that you should choose questions not only on the basis of consequences, but also on whether they have answers and whether you have thought through what you will do with the answers.

I realize that I have left a good deal of mystery in your mind about indeterminacy. This mystery arises naturally from our wish to ask questions without first asking ourselves whether the question makes sense and should have an answer, and what we are going to do with the answer if we get it. Avoiding this mystery is the third application of quantum mechanics, the selection of appropriate questions.

It is, as we have seen, useless to ask the precise time of an atom's transition from one precise energy state to another. But we can reformulate the question and ask: In a collection of atoms excited in a specified way, how much of what kind of energy will be emitted per unit time? To that question we get a precise and useful answer, one we can compare with experiment. We would not have known what to do with the answer to the earlier question even if we could have found it. Although information about an individual atom is constrained by the Indeterminacy Principle, a collection of atoms of observable size yields precise answers to any question. No uncertainty here.

The packet of energy emitted by an atom has a quantized energy and is emitted or absorbed as an indivisible unit. It thus seems to be a particle, and we call it a photon. But in moving from place to place, its motion is as if it were a wave, with a definite wavelength and the defining characteristics of a wave, namely interference and diffraction. Is it a particle or a wave? There is no answer to this question, unless you say that it is a "wave-particle," which simply begs the question.

The trouble is that I have not asked an appropriate, useful question, however reasonable the question may have seemed before I understood what was going on. What would I have done with the answer? If the answer had been "wave," I would still have to have dealt with its birth and death as a particle; if the answer had been "particle," I would still have to have dealt with its motion from place to place as a wave.

A "research assistant" is a graduate student who is paid (usually from an external grant) to conduct research under the supervision of a professor (who usually is the person who obtained the grant). The work of the "RA" usually will be used as part of the thesis he or she submits in order to obtain a doctoral degree. The Internal Revenue Service insists that the RA must be either an employee or a student, and of course, the IRS comes down on the side of taxing him or her as an employee. But there can be no answer when the question is put as either, or. The IRS ought ultimately to understand that we have here a *joint* product, mostly educational, and a product that ought to be encouraged by treating the pay as one would that from a fellowship. A similar situation occurs with residents or interns in a teaching hospital. Far more consequential examples occur frequently in corporations, as in the fixed costs of maintaining a corporate aircraft, corporate library, or corporate lunch room. An attempt to assign costs of such enterprises too precisely to corporate accounts is quite likely to kill the activities. The executive must decide on more general grounds than on the worth to individual corporate units whether the functions are worth the cost.

There is a wide variety of questions that do not have answers. You must learn that it is useless and frustrating to ask such questions and that you should think before asking any question. In bad weather, you are required to hover around an airline ticket counter to pick up any real news the instant it arrives. But it is positively embarrassing to be hovering there and to have to listen to your fellow travelers asking unanswerable questions.

Only the very young believe that every question has an answer and every problem has a solution. Often one must restate the question and reformulate the problem before a useful answer or a solution can be obtained. Often the key thought process will be thinking about what you will do with the answer. Asking questions when you have no clear idea of how you will use the answer is not only unproductive but can be damaging.

Every question you ask gives a message to the troops. If your understanding of the area of questioning is inadequate or if you do not have in mind a constructive use for the answer, the message may be wrong or at the very least unintended. Asking the wrong question would be a bad mistake.

In this chapter, you have seen some of the fruits of applying the ideas of quantum mechanics. The first of these is the quantization, the collecting into appropriately sized packages, of management decisions. The second is the interaction between the process of measurement and the quantity being measured. The third is the choice of appropriate questions and the useless frustration of asking unanswerable questions. In passing, I hope I have persuaded you *not* to blame uncertainty, wherever encountered, on quantum mechanics, which is a very precise science. Uncertainty in human affairs usually owes its origin to the *entropy* that we shall examine in Chapter 17.

5

Visual Management

Visualization dominates scientists' thought and communication. If you join a pair of scientists at lunch, the chances are one of them will sketch on a napkin (assumed to be paper—no fat cats here) the results of his or her most recent experiment or theory. Most probably this visualization will be by a *graph*, and the graph will be a living, working sketch, quickly marked up to represent what-ifs, possible errors, and alternate explanations.

Communication by graphs is a powerful tool and has become even more common with the arrival of the personal computer. But the underlying, even more powerful use of graphs is for the visualization of a relation between (or even among) variables, even if the originator never shows the graph to anyone or perhaps has it only in his head and never reduces it to paper or to a computer screen. Graphs are a mode of *thinking*, as well as communicating.

Before going any further, I pause to deal with two problems of nomenclature. First, the relation conveyed by a graph, if sufficiently simple, can frequently alternatively be conveyed by a bar chart (Fig. 5-1). Bar charts are cruder and more limited in application. They are used primarily when some quantity is being portrayed as a function of time. Graphs are more versatile.

Second, it is traditional to plot the independent variable along the horizontal axis (it is called the "abscissa," but we shall not use that term) and the dependent variable along the vertical axis (the "ordinate"). Admittedly it is sometimes not clear which

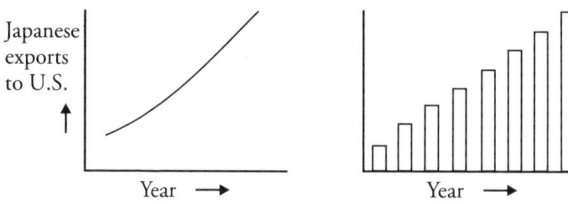

Figure 5-1. A graph is at the left. A bar chart is at the right.

of the two related variables is the independent one. But it is clear enough if one of them is the time; time is quintessentially independent: "Time and tide wait for no man."

In many cases the decision as to which of the two variables is independent may be the vital decision. This decision, of course, is far more important than just a drafting nicety. Identification of cause, plotted horizontally, and effect, plotted vertically, can be the key to understanding phenomena.

Sometimes the two variables are on equal terms, as in lines on a map, a graphical representation of a path on terrain.

Now I must introduce some nomenclature. In the graphs below, y and x can be anything you wish. For example, y can be the total cost of a shipment of items and x can be the number of items in the shipment. In the plots that follow we can continue to think of this example.

The first plot, at the left in Fig. 5-2, shows the situation where y is *proportional* to x. Otherwise stated, y is a constant multiplied by x. This is the somewhat idealized situation in which the cost per item is absolutely independent of the number of items shipped.

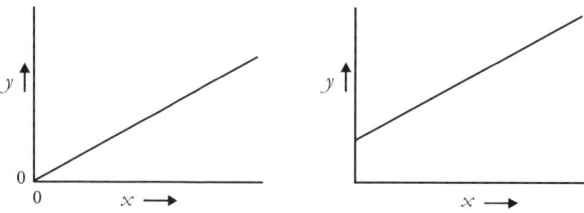

Figure 5-2. Proportional behavior is at the left. Linear behavior is at the right.

The second plot (at the right in Fig. 5-2) shows a situation where *y* is a *linear function* of *x*. Otherwise stated, *y* is a constant plus another constant times *x*. In our example, there would be some cost of the shipment of an empty truck or empty boxes when the total shipment contained zero items; therefore *y* is not simply proportional to *x* but is a linear function of *x*. The name "linear" arises, of course, because the graph is a straight line. This second plot might also apply to the cost to produce a product as a function of the number produced. Even before the first one is produced, there is the cost of development and of the putting in place of promotion and distribution channels.

The next two graphs (Fig. 5-3) show *nonlinear* behavior. In the plot on the left *y* increases faster than linearly with increasing *x*. On the right, *y* increases more slowly than linearly; it shows saturation behavior.

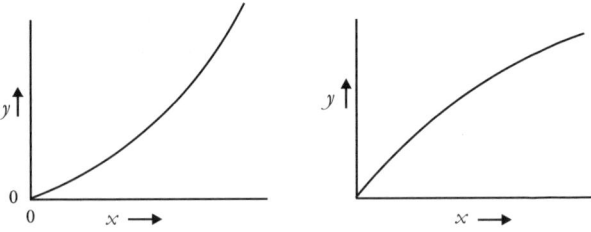

Figure 5-3. Nonlinear behavior.

Graphs are especially useful for visualizing and communicating motion or changes with time, and a language has developed for some characteristic dependences.

A *step function* is illustrated first, an abrupt change (either upward or downward) in a quantity that elsewhere is either constant or slowly varying (Fig. 5-4).

This might be, for example, interest cost as a function of time when an abrupt increase was caused by borrowing to finance an acquisition. Note that this is an example of *dis*continuous behavior; we spoke of continuous and discontinuous behavior in the preceding chapter.

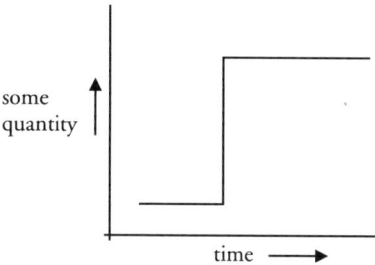

Figure 5-4. Step function.

A *pulse* is an abrupt (again either up or down) step followed by a reversal (Fig. 5-5) to nearly the original value. A division manager may argue that he needs a pulse of resources (budget dollars or head count) to accomplish an agreed upon goal. But a step function would make life easier for him since he would not have to contract his operation after expanding it. Corporate management must watch carefully to make sure he does not surreptitiously get that step function when all that was agreed to was a pulse.

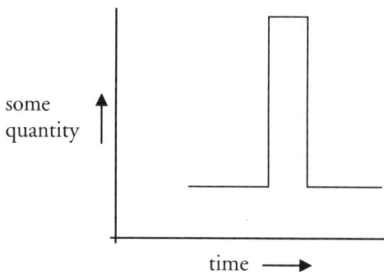

Figure 5-5. Pulse.

You must always be on the lookout for situations in which a pulse of additional resources can profitably be injected into your operations. Suppose, for example, that you identify a queue, either of orders waiting to be processed, parts waiting to be assembled, or of people standing in line. Your division manager looks upon this as a reason for increasing his staff. But if the queue is not becoming longer, you do not need a step function of additional employees; the rate of satisfaction at the counter is just as great as the rate at which people are added to the queue. A temporary injection (pulse) of activity will reduce the queue to zero, serving everyone better, and yet will have little impact on continuing costs. You have probably seen this happen when an alert manager at a fast-food restaurant adds help (temporarily!) to the longest queue.

Another frequently encountered graph shape is a *sawtooth* (Fig. 5-6). In real life the teeth are not symmetrical, and this basic shape may sometimes be superimposed on a curve (rather than continuing constantly to the right as shown). A common example is corporate office head count or overhead, which tends to creep upward until a Draconian effort sharply reduces it, only to launch a new upward creep.

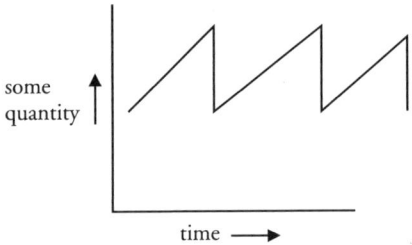

Figure 5-6. Saw tooth.

Another example from my experience is overclassification in the U.S. Department of Defense. Year after year more items are labeled "Secret" or "Top Secret" or even "Code Word, Special Access"; then an abrupt cut in classification occurs, usually by an alert and aggressive new official, only to have the excess secrecy begin again its slow, steady climb.

Of course, the sawtooth can face the other way. When someone in the audience asks a soft-spoken speaker to "speak up," the sound level will usually rise abruptly but then slowly settle back down again. An added example is the way an abrupt increase in attention to fire safety is produced by a fire drill; after each drill, attention declines slowly until the next drill.

Another common example is the *growth curve* (Fig. 5-7). Typical examples are the numbers of a new product that are sold each year after their first appearance in the market, or the utilization (number of customers per year) of some new service.

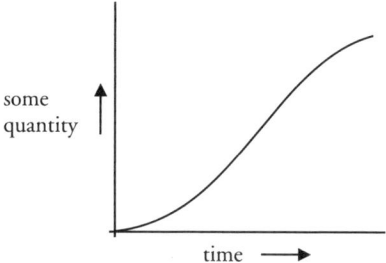

Figure 5-7. Growth curve.

The *learning curve* (Fig. 5-8) typically represents the reduction in cost per unit as manufacturing experience accumulates. For a while, the efficiency of producing each new unit increases by applying what has been learned on earlier units to the machinery and training on the production line.

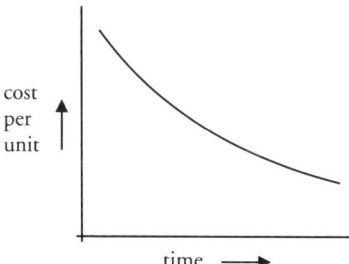

Figure 5-8. Learning curve.

But eventually all the experience has been exploited and the cost per unit approaches a constant value, essentially irreducible unless major changes are made in the design or the process.

Another important shape is the *exponential increase* (Fig. 5-9). This curve, which could, for example, be the principal value of an investment yielding a constant, compound interest, is characterized by the fact that its *fractional* increase is the same for each increment along the horizontal axis. Another example is the total number of television sets or videocassette recorders in use as a function of time, soon after their introduction. When television sets or video cassette recorders were introduced, the number sold per year was about proportional to the number already in use (the people who had sets told their friends); the fractional increase was thus constant, and the early part of the growth curve was an exponential increase.

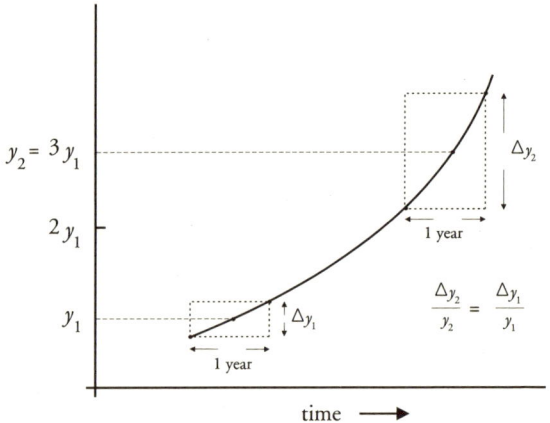

Figure 5-9. Exponential increase.

(You may wish to skip this paragraph, since it is more detailed than most of this book.) The constancy of the fractional increase can be expressed in symbols as follows: Let the increment of y (the variable along the vertical axis) be called Δy corresponding to a constant increment Δx (such as one year) in x. Then this curve is such that $\Delta y \div y$ is the same at all values of x and y. An example of such a curve is $y = 10^x$ and hence the label "exponential." More generally, the equation of such curves is $P = P_o R^x$, where P_o is the initial value of P, the value when $x = 0$.

For example, P could be the principal value of an investment that started at P_o = \$10000 and that earned 8% per year, all of which was plowed back into the principal. Then R = 1.08 and x is the time in years. You will remember from the preceding chapter that this can also be written $\log P = \log P_o + x\log 1.08$, and so the logarithm of P is a linear function of x (i.e., a straight line when plotted as a function of x; see Fig. 5-10). As in Chapter 3, you may find it difficult to think about this just by reading; pencil-and-paper examples and sketches will help enormously.

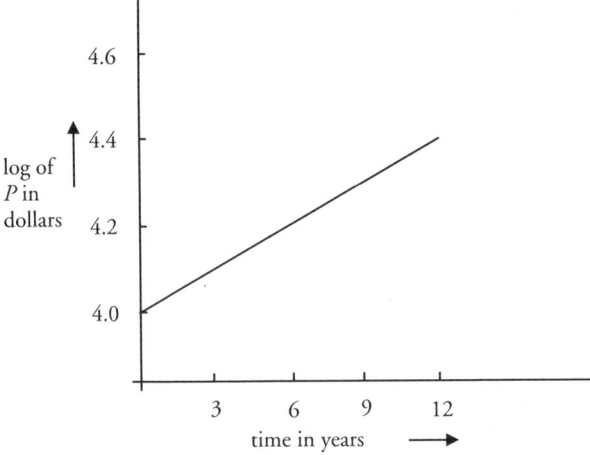

Figure 5-10. Exponential increase; logarithm of the quantity is plotted.

For this function, or for functions that are continuously growing even if they do not exactly fit this definition, it is a great convenience to distort the graph paper so that distance (in inches) along the vertical axis is proportional to the *logarithm* of the number plotted. The new paper, Fig. 5-11, is therefore distinguished by the fact that a given increment of distance on the vertical scale always equals the same *fractional* change; for example, the distance from 1.0 to 1.1 is the same as from 10 to 11, both representing the fractional change of a factor of 1.1 or an increase of 10%.

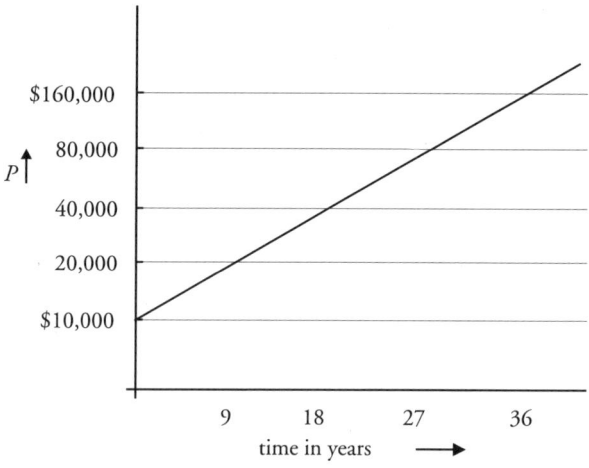

Figure 5-11. "Semilog" graph paper.

On this "semilog" ("semi" because only one axis is logarithmic) paper, our function above is a straight line since the logarithm of P is being plotted. Plotting on semilog paper makes it easier to extract growth rates from data, as we shall see in a few pages. (Note, incidentally, how the Rule of 72 works here: 72 ÷ 8 = 9, and the value of the principal plus its compound interest at 8% doubles in 9 years.)

It is remarkable, and perhaps testimony to the "two cultures" of C. P. Snow, that the semilog graph, although used in technical articles by economists, did not enter the securities industry until the late 1930s, and even then was viewed with suspicion, as if it were a confidence man's deception. From such graphs it is easy to determine approximately the fractional change in the price of a stock year after year, and when this is added to the dividend it gives us the *total return,* which is of course the only "income" that counts in the long run.

I promised in Chapter 3 to return to the telephone lines and the db business and make it a little clearer through graphs. Plotted in Fig. 5-12 is the signal power on a long telephone line with the characteristics cited in Chapter 3. The upper plot is on ordinary graph paper, which quickly ceases to portray much after the signal has been attenuated by a factor of 10 or 100. The lower, of the same situation, is on semilog graph paper, which continues to be useful throughout.

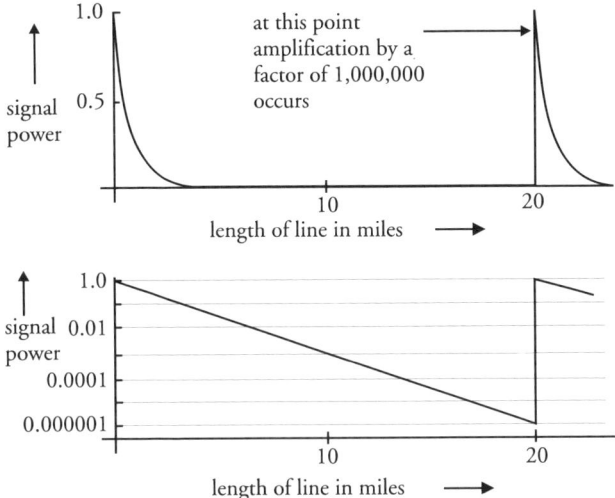

Figure 5-12. Signal strength on a long telephone line. The upper plot uses ordinary graph paper. The lower plot is on semilog paper.

Toward the end of Chapter 12, we shall examine another kind of plot, a *scatter diagram*, and illustrate it in Fig. 12-11. These diagrams are useful as ways to learn relations between variables. For example, order date can be plotted vertically and shipping date horizontally; examination of the scattered points can reveal patterns in backlogs. Similarly, pricing margins (percentages) can be plotted against unit prices or against unit volumes or against dollar volumes in order to reveal actual pricing behavior patterns.

Before going on I should warn you about "suppressed zeros." You will often find that either the horizontal or the vertical scale does not start from zero. Now this may be for a good reason and be perfectly defensible. For example, if the horizontal scale is time in years, it can hardly start with the origin of the universe and any other zero would be arbitrary; we therefore start with whatever year is convenient for the data or problem at hand. Also, all of the known data may be for a restricted range of values of a function, say, between 112 and 137; it then is appropriate to let that scale run from 110 to 140 to spread out the data in more readable form.

But zeros can be casually suppressed or even suppressed intentionally to mislead. If the data run from, say, 6 to 10, starting at 5 exaggerates the variation and gives a quite misleading visual impression. Consider the accompanying chart (Fig. 5-13) from a federal government report: The viewer gets and is likely to remember the impression (unless he studies carefully the vertical scale) that in 1991 deaths in childbirth have decreased almost to zero. But the actual ratio from the curve of infant mortality, for example, is only a factor of 3 in thirty years. There was no reason to suppress the zero here (the vertical scale could just as well have started with zero), which was doubtless done casually; the wickedness comes when it is done intentionally, usually to exaggerate small changes or insignificant differences. (There is another, less common, defect in this plot: Note the wholly unnecessary distortion of the horizontal scale.)

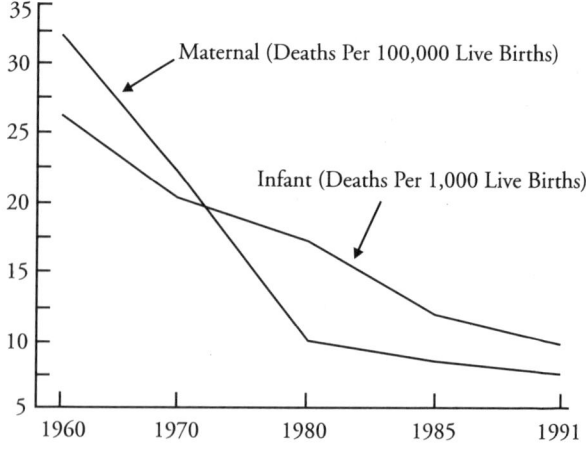

Figure 5-13. Maternal and infant mortality rates. Note suppressed zero and distorted horizontal scale.

Another warning is in order about graphs that are generated not from data, but from concepts or ideas buttressed to some degree by data. In an earlier, more primitive state of economics, one frequently saw curves that had no basis in fact but from which conclusions were drawn nevertheless. For example, an economics paper printed the graph in Fig. 5-14 and drew its principal conclusion from the intersection of the line and the curve (other than the intersection at 0, 0). But there was no reason presented why the curve had this shape rather than the curves in Fig. 5-15 that would not produce an intersection.

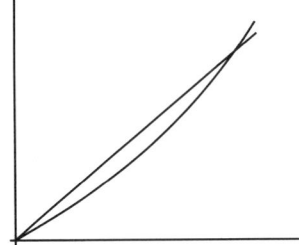

Figure 5-14. An assumed relation between economic variables.

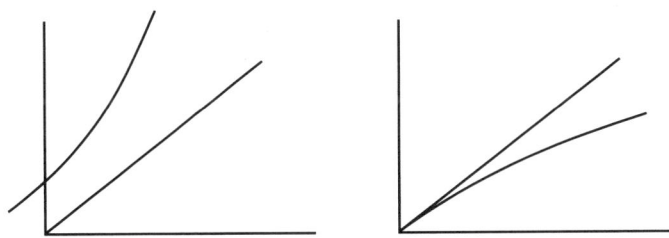

Figure 5-15. But the relation could look like one of these.

Another example is the "Laffer Curve," which had high visibility in the late 1960s. It was a plot of the total revenue from an income tax as a function of the tax rate, from 0% to 100%. Clearly at 0% there is no revenue, and presumably at 100% there is no revenue, since why should anyone organize his affairs in such a way that all of his income was taken away? But those two points were the *only* known points. The curve was often drawn like the graph in Fig. 5-16, which would say that above a 50% rate one actually obtained *less* total revenue as he increased the rate. It was even drawn sometimes as in Fig. 5-17, where rates above about 30% would be "counterproductive" (to use a wretched Washington neologism).

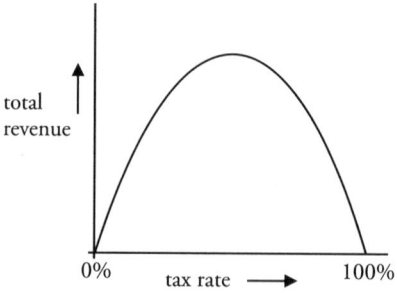

Figure 5-16. Revenue as a function of tax rate.

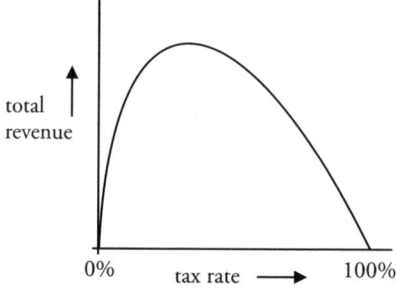

Figure 5-17. "Laffer Curve."

But what if the curve was like the one in Fig. 5-18, which has just as appealing speculation behind it? With this curve, total revenue continues to increase well above the 50% rate. (I am not including here, of course, the damaging effects of higher rates on investment, innovation, and the economy.)

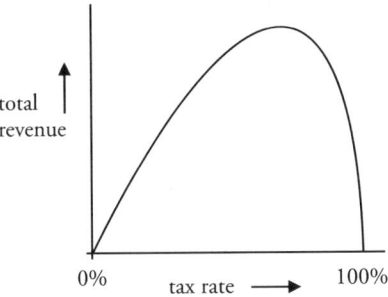

Figure 5-18. Possible alternatives to Laffer Curve.

Another warning is in order about *curve fitting*. Suppose we can measure the productivity p of hourly workers assembling a particular product and plot it as a function of time (Fig. 5-19). This p may be, for example, the number of units assembled per eight-hour shift. It is tempting to draw a line, more or less through the points.

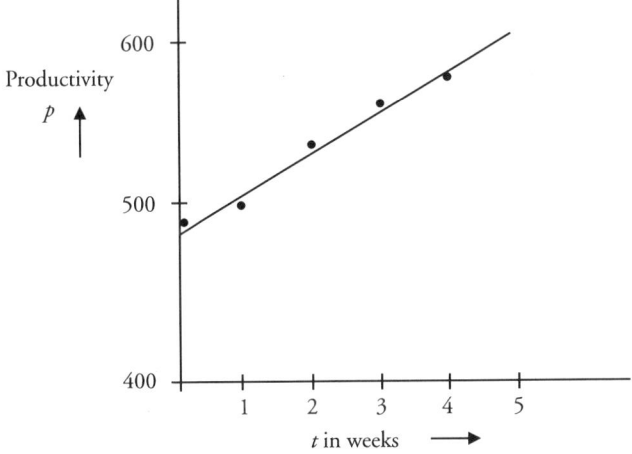

Figure 5-19. Productivity for the first few weeks.

Now there are sophisticated ways of teasing out the equation of that line; "least squares" is the most common, in which the line is determined such that the sum of the squares of the deviations of the data points from the line is minimized.

But eyeballing is good enough here and it usually is actually to be preferred to more elegant methods, since it does not mislead us into conclusions from inaccurate data more or less fitted by a line. The line $p = 470 + 27t$ looks pretty good, does it not? If you have not done much of this, I suggest that you try putting in some values of t and calculate the related p. At $t = 0$, $p = 470 + 0 = 470$; at $t = 3$, $p = 470 + 81 = 551$; and so on. The "470" and "27" are called "parameters"; we have fitted the data using two parameters (sometimes in the functional technical illiteracy common in bureaucratic circles these will be called, quite wrongly, "perimeters"). From the parameters we see that over this limited period the productivity increases are about 27 each week or $27 \div \sim510 = \sim5\%$ per week; any more precise conclusion is clearly unjustified.

You will have anticipated that complications will enter if productivity increases continue over many weeks, as in Fig. 5-20.

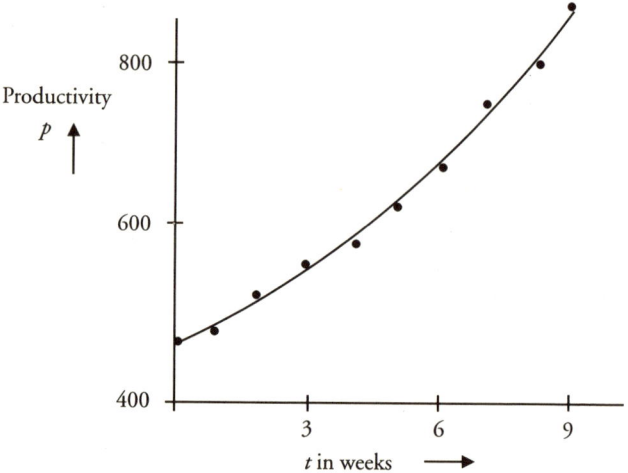

Figure 5-20. Productivity over a longer period.

(Caution! I have suppressed the zero, but for good reasons.) We are tempted to fit these data with a curve, and we can do so (for a while!) with $p = 470 + 22t + 2t^2$; we have added another parameter, chosen to fit the data. We can always obtain a closer fit by using more parameters, but it is very treacherous to assign *meaning* to the parameters derived in such an exercise. (It has been said that "with six parameters one can fit an elephant"; of course this is something of an exaggeration, but the point is that none of the parameters found in this way would have any significance.)

Rather than blindly adding parameters, however, we should note that the upward curving trend looks like the exponential encountered in Fig. 5-9, and this resemblance suggests that we try a logarithmic scale (Fig. 5-21). Eyeballing these data points lets us draw a line. The equation of this line is $p = 470 (1.05)^t$, and we did not need any mathematics to find it. All we needed to do was to note that the productivity doubled in about 14 weeks and to apply the Rule of 72: $72 \div {\sim}14 = {\sim}5\%$; thus at the end of one week ($t = 1$), $p = 470 \times 1.05$; at the end of two weeks, $p = 470 \times (1.05)^2$, and so on.

When plotted this way it is easy to detect *deviations* from *exponential growth*, as they will reveal themselves as changes in slope (a different exponent), step functions (e.g., if new machinery made a step increase in productivity), or other changes. You well know that these productivity increases will not go on indefinitely; eventually the curve will cease to rise exponentially and will become the growth curve we encountered earlier in this chapter (Fig. 5-7).

I have noted the temptation to enlarge the number of parameters to get a more impressive fit and the danger of assigning meaning to them when they have been chosen solely to fit a curve to data. A particularly treacherous use of curve fitting keeps popping up in the investment advising industry. A particular period (say, 1975–1995) and a particular set of prices (say, the Dow Jones Industrial Average) are selected. A curve is then fitted to the data points of prices as a function of time. This fit to the curve can have as many parameters as patience and time on a computer permit; it is easy to fit with twenty or even forty, and such a fit will reproduce in amazing detail a near

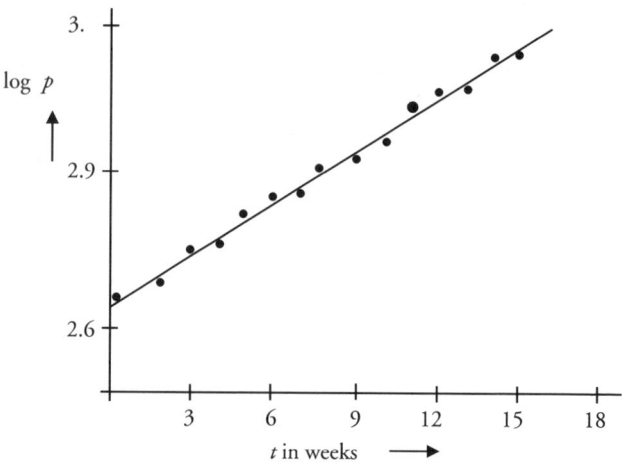

Figure 5-21. Fitting a curve to exponential growth.

approximation to the data points. (There are many ways of doing this; the most popular with scientists is Fourier series, which is not difficult but would carry us too far afield to pause here to explain it.) Even if only a few parameters are used, the basic up-and-down moves are reasonably faithfully reproduced.

So far so good, but the next (wicked) step is to assert a "theory" that these parameters will *predict* the performance in the *next* ten years. Of course there is no reason at all to believe that they have any predictive capacity. And the final (criminal, it should be) step is to "test" the "theory" by "how it would have worked in the period 1975 to 1995." Of course it "works" wondrously! Curve fitting, if not properly exposed, can lead directly to bankruptcy. Representation and communication are one thing; *prediction* is totally different.

Toward the end of this chapter, I have included some warnings about the *mis*use of graphs, warnings that may level the playing field between you and anyone playing fast and loose with graphical representation and especially with conclusions from graphs.

But the major message of this chapter is that graphs, even if only in one's head, are a powerful aid in thinking about the relations between or among variables. Behavior of all kinds, especially business data as a function of time, can be represented by characteristic shapes like step functions or saw teeth. Augmented by a few tricks and labels, graphs provide a quick and effective means of extracting the meaning from data and communicating it to others. An especially useful trick is to plot a function on semilog paper in order quickly and easily to estimate growth.

6

Direction

In common usage, "direction" begins with the literal meaning of the orientation of motion on the earth and proceeds to the metaphorical meaning of control or guidance, motion in a more complicated space than the real, geographic space. You may think that with "outer space" and "space stations" we already have at least one space too much, but I seek here to persuade you that creating new spaces, at least in your head, is a powerful and productive mode of thought and explanation. This approach is especially useful by an executive who must incorporate into his thinking and direction the differing goals, ambitions, and capabilities of his associates.

We start with a familiar space, a *map* (or "chart" if the territory is wet), in order to get a firm grip on the process. On a map, oriented in the usual way with north at the top, both the horizontal and vertical scales are marked off in miles, and this is "real space," albeit somewhat idealized since it is the projection on a plane of the undulating three-dimensional surface (we shall have more to say about "projections" later).

A motion from A to B (see Fig. 6-1) can be described by a *vector*, a line joining B to A and with a pointer indicating that the motion is directed toward B. The length of the vector is the distance traveled, and the orientation of the vector on the map gives the direction traveled.

We can add vectors (Fig. 6-2); if subsequent to going from A

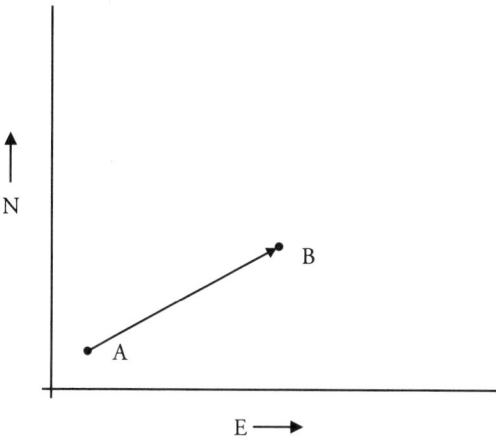

Figure 6-1. A vector.

to B we were to go on to C, the new leg of our trip is the vector BC, and the total trip is the sum of AB and BC, which is AC. This is obvious enough on the map.

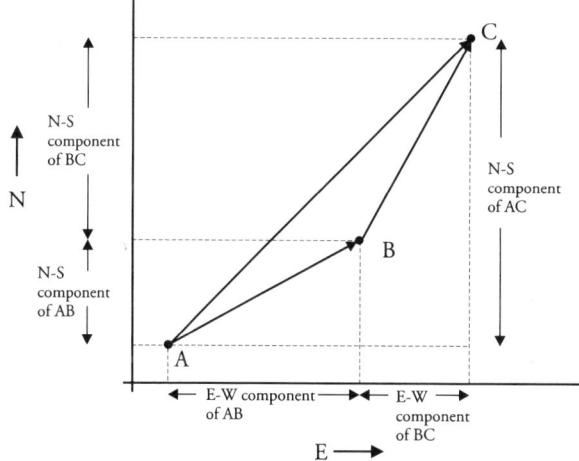

Figure 6-2. Addition of two vectors.

But if we wish to get a quantitative, numerical description of the vector sum, it is necessary to break each vector down into its *components*, which are the *projections* of the vectors on the east-west and north-south axes. These projections are made by drawing the dashed lines at right angles to the coordinate axes. Then, as can be seen on the map, we add the E-W component of AB to the E-W component of BC to get the E-W component of the result, and similarly for the N-S components. These projections are simple numbers ("scalars") and can be added and subtracted as such.

We can make a single step of generalization of this representation by creating a new space, "velocity space," in which both the speed and the direction of motion are portrayed (Fig. 6-3). Sailors are accustomed to putting these velocity vectors right on the chart, since the N and E directions are the same in either space, but they carefully recognize that the units (miles and miles per hour) are different. I have explicitly introduced velocity space, however, since not everyone is a sailor and since this makes the next step in generalization easier to grasp.

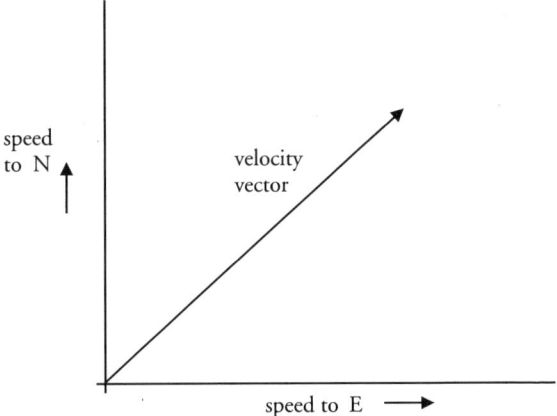

Figure 6-3. Velocity space.

This next step is a grand step, since to make it I must abandon the comfortable arena of maps and go into a more abstract space. I make this step by describing a particular example, a "decision space."

Consider a physics faculty meeting that is attempting to assess (and later to represent to the dean and the president) the will of the faculty in response to a projected budget opportunity; for example, next year's budget may increase 5%, and that coupled with two retirements and one defection permits 20% of this year's budget to be applied in ways to be chosen. The big issue (although there are others) is: How much do we apply to hiring new faculty (replacements plus possible expansion) and how much do we apply to increasing faculty salaries? We draw a decision space Fig. 6-4 in which the axes represent not distances or velocities but the emphasis to be placed on the conflicting uses of money. All realize that they must balance the two aims, that they cannot "go all out" for both size and salary scale.

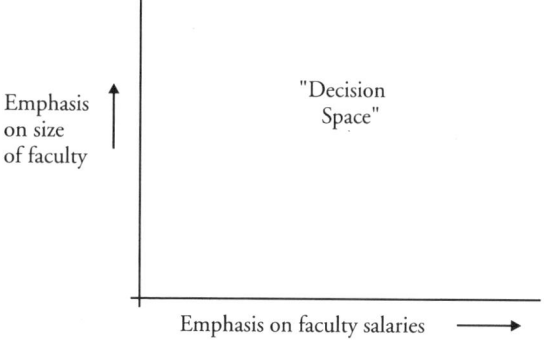

Figure 6-4. Decision space.

Professor A in Fig. 6-5 is most worried about teaching loads, and his thrust is strongly for size and only modestly for salaries. Professor B feels so strongly about salaries that she would even decrease the size of the faculty. Professor C is asleep. Professor D "can see both sides" and his apathy has arrived.

These same choices can equally well be represented by vectors (Fig. 6-6). The vector representation is perhaps a little more intuitive, since the direction of each arrow is the direction of each professor's thrust and the length is proportional to the intensity of his conviction (or vehemence of his opinion).

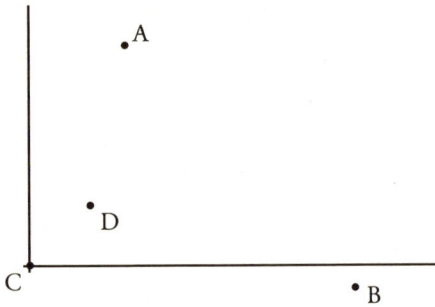

Figure 6-5. Decision space with four preferences.

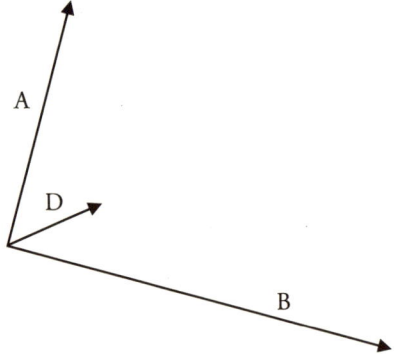

Figure 6-6. Preferences as vectors.

To get a representation of the will of the total faculty we must add these vectors (Fig. 6-7). We do this just as we did for vectors in "real" (map) space, namely by adding components.

We next add the vectors of all 32 faculty members. We have not here specified how the other 28 think, and so we will not do the addition explicitly here. The addition yields the resultant R; perhaps it comes out rather strongly for raising salaries and nearly as strongly for expansion, and that is what is plotted in Fig. 6-8. (This R is now plotted to a different scale since it is likely to be much larger than the thrust of a single member.)

I am not suggesting that this process is reduced to paper; as the chairman listens to the opinions, he is performing this vector addition in his head. He doubtless adds his own opinion,

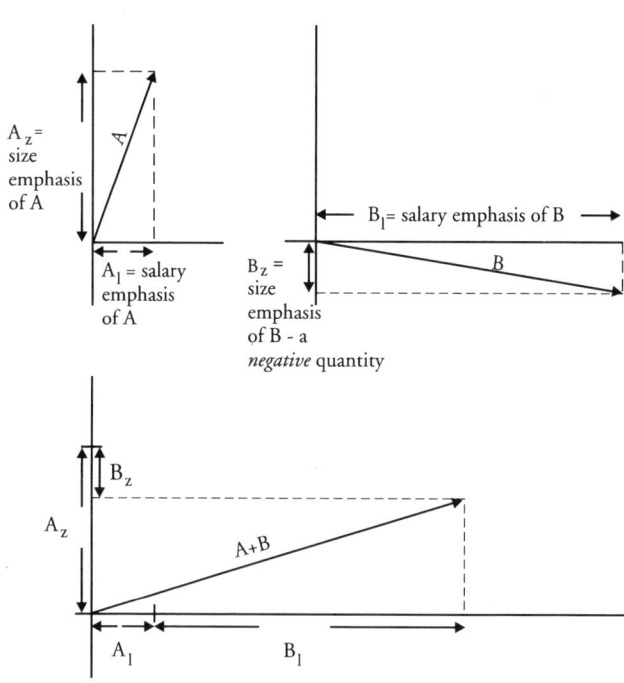

Figure 6-7. Upper part, taking components; lower part, adding components.

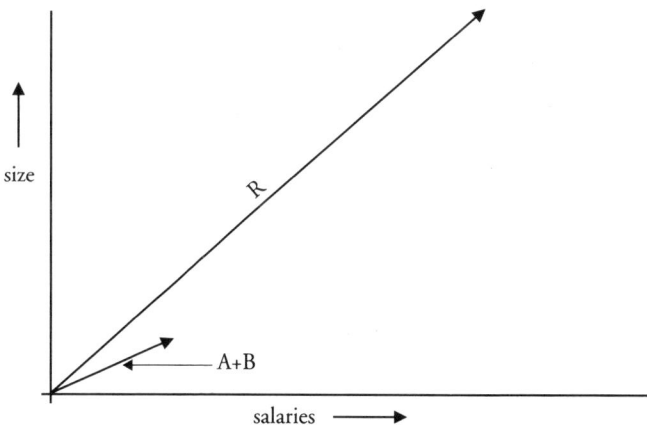

Figure 6-8. Resultant *R* of all preferences.

and he also is likely to weight the opinions of the others differently, multiplying each vector by a weighting factor giving effect to his confidence in the judgment of that professor and in the professor's commitment to the institution. There is thus room for leadership by a skilled and fair chairman, and a valuable representation of faculty interests has been obtained and can be communicated to others. If he or she is fair and honest, he will represent the will of the faculty *without* his weighting and then separately *with* his weighting. (I do not mean to imply that this is the only leadership the chairman provides; other areas, such as the direction and goals of the department and evaluation of the performance of individuals are more important. And I am certainly not advising you to run your company in this way.)

Most organizations require at least some teamwork. If each individual's goals and activity are represented by vectors, then behavior as a team means that to a large extent the vectors must point in the same direction. A vector at right angles to the general direction of the team contributes nothing to team performance, and a vector that is even slightly pointed in the opposite direction detracts. This standard approach must be tempered, however, by recognizing the value of listening to the "person who thinks otherwise"; we shall encounter such persons in the next chapter.

There is much consideration in the management literature and in the press of letting the production workers redesign the manufacture of a product. The aim is to take advantage of the experience of those actually doing the work to cut time, rejected parts, and scrap, to increase quality, and to improve the working environment. I cannot advise you as to when (or whether) to use this technique, but I do suggest that *if* you use it, you harvest the results by the process just described. Using this process lets you profit from all of the contributors, often with conflicting suggestions or conflicting calls on resources. As I have said, it may or may not be useful to reduce the result to paper. Note that it is easy to consider a third dimension (e.g., hiring young vs, hiring experienced candidates for open positions), although it is not so easy to plot on two-dimensional paper. In fact, one can *think* in any number of dimensions, and it is thinking in these terms that counts, not putting them on paper. You see that not only are you now able to deal with "the fourth dimension," but you can take *many* dimensions in your stride.

A related way of thinking in decision space is to consider pairs of alternates. For example, the salary-size choice can be thought of as positions along a line (Fig. 6-9). Each participant feels a *tension* along this line between the choices and registers the strength and direction of his view by his position on this line. Although something has been lost (C and D register the same), the basic conclusion has not been altered when the chairman reports the result of all these vectors. But something has been gained, since explicitly acknowledging the tension between conflicting choices contributes to our thinking.

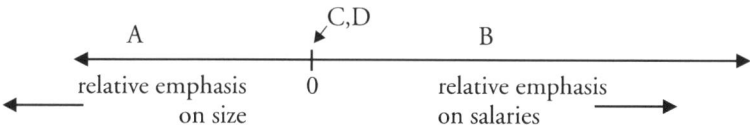

Figure 6-9. Portrayal by vectors along a line.

We can, of course, think of tensions along a large number of axes; in our example, they might be salary vs. size, young vs. experienced, experimental vs. theoretical specialty, heavily research vs. heavily undergraduate classroom-teaching orientation, and so on. The sum of the projections of each faculty member's thrust on each of these axes provides the total opinion of the faculty. Each of these conflicts has been represented as a tension along an axis, a one-dimensional representation; but by considering each dimension in turn, we have captured the subtlety and complexity of the multidimensional question.

Business examples of teasing out the thoughts and priorities of managers will probably come quickly to mind. In the design of a new product, for example, tension occurs in many dimensions: Quality contends with cost; limiting weight contends with durability; use of latest technology contends with confidence in predicting useful life; features that make the product more capable or attractive contend with simplicity to make it easier for the customer to understand how to install and use it; and many others.

Similarly, this mode of thinking can clarify tensions in the design of a new promotional campaign for a product or service: television advertising vs. print advertising, advertising vs. introductory pricing, targeting older customers vs. targeting younger, pretesting in a limited area vs. rapid introduction worldwide, claims of capability vs. warranty liability, and many others.

Letting the "troops" have some input in deciding the location of corporate headquarters can proceed just like the faculty example cited earlier. And the design of the staffing of a human resources department can profitably include in a similar way the demands and priorities of the managers whose operations will be served.

In fact, practically *all* business decisions can use this mode of thinking because they are *constrained*: Some function or feature must be limited because its unconstrained development would interfere with another valuable function or feature. If no other constraint occurs, there is always cost, but time available, experienced manpower available, capital investment required, and other constraints are also important.

Much media reporting is narrowly *one*-dimensional. Perhaps the most common example is the practice of describing every individual by his projection on the "conservative"-"liberal" axis. But to understand an individual in the complicated modern world, one must consider many, many dimensions. And the conservative-liberal axis must be refined and separated into many axes, each for tension on a particular issue. Business decisions invariably involve tensions in many dimensions, but the press will just as invariably portray a limited number (frequently only one) of these, for example, bottom-line earnings vs. concern for the environment.

I have rather casually introduced the concept of the projection of a vector on an axis. The concept is even more useful when generalized. We can think of the projection of one vector on another vector. If these vectors are the thrusts in decision space, then the projection (Fig. 6-10) clearly represents the amount of A's thrust that agrees with the direction of B's thrust.

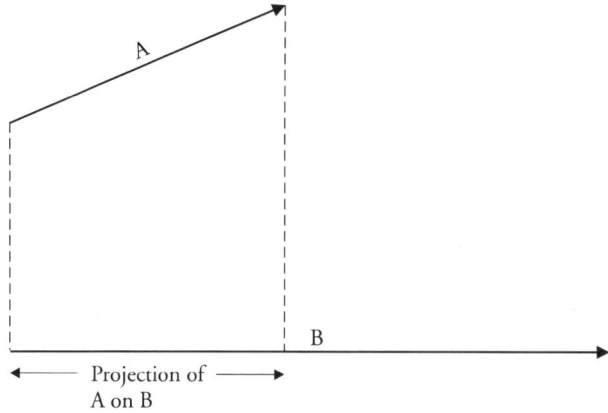

Figure 6-10. Projection of one vector on another.

Total agreement in thrust, though not necessarily to the same extent of conviction or vehemence, occurs, at the left in Fig. 6-11, when the vectors are parallel and the projection of A is as long as A itself. Total disagreement, at the right in Fig. 6-11, is similar.

Figure 6-11. Vector representation of agreement and disagreement.

The projection of B on A is, as you can see from the diagrams, not the same as the projection of A on B.

Consider (Fig. 6-12) a vector H as the promise of a husband's career measured in any terms you like (satisfaction, service, salary, respect, security, etc.); similarly, the vector W is the promise of the wife's career. These careers will in general have different directions (nature of job, location, hours, etc.). In the example portrayed here, the wife's career is more promising than the husband's. The wife helps the husband to the extent her career moves in the same direction as the husband (e.g., can be pursued in the same city, needs the same reference books or computer software, etc.). This is shown at left in Fig. 6-13. Similarly, the husband helps the wife (at right in the illustration).

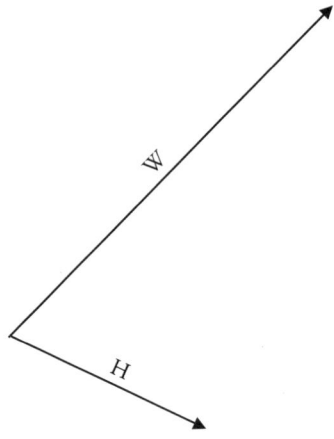

Figure 6-12. Graphical representation of career promises.

Of course we must think of *all* the dimensions, and of course there may be in some of them head-on conflict. For example (Fig. 6-14), in the geographical dimension, the wife must work in Paris and the husband cannot bear to be away from the Charles River.

You will readily appreciate that since these examples involve human preferences and performances they are necessarily only

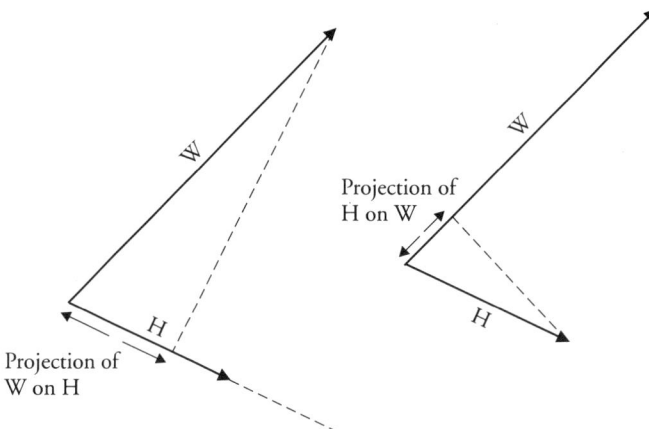

Figure 6-13. Left, wife helps husband. Right, husband helps wife.

Figure 6-14. Conflict situation.

crudely quantitative, but the concept of projections in many dimensions is a powerful mode of thought.

The special case where the vectors are at right angles merits particular attention. We then say that they are *orthogonal*. That is a perfectly acceptable English word meaning "at right angles," but it is rarely used in general conversation. It deserves wider use. In our vector-projection mode of thinking, orthogonality becomes an important concept. Two careers, decisions, courses of conduct, or paths of business development are orthogonal if neither helps nor hinders the other. An extremely influential person can be totally useless if his or her influence is orthogonal to the important issue at hand. A product line can be so nearly orthogonal to all the others that divestiture is a realistic possibility.

Identifying orthogonality can simplify business decisions; if the development of product A is truly orthogonal to the development of product B (not using the same pot of resources or talent), then the decision to undertake one or the other can be made in isolation. But any deviation from orthogonality must be recognized carefully, to avoid unintentionally clobbering one development by the pursuit of the other.

For more than half of my lifetime, I have needed to tease out from letters of recommendation all of the wisdom I could about the recommended individual. Many times the artfully chosen language, when minutely studied, produced not a recommendation at all, but an orthogonal comment. At these times, in order to preserve whatever good humor I was capable of, I thought of the old story of the kindhearted woman who was determined to rid herself of her terrible cook. She finally found the cook another job, provided that she would write a letter of recommendation. She wrote: "Mrs. A has been my cook for 11 years." No problem with that. Then: "I am very happy to write a letter for her." Disingenuous, but true. But she could not stop there and simply must find a third sentence or she would be stuck with the cook. Finally, she wrote: "A fine meal from this cook is a rare treat."

On my first technical job, I became overly excited in my presentation to an experienced engineer. His put-down of my "discovery" was memorable: "True, but not very interesting." He was acknowledging a certain positive projection on the false-true axis, but asserting (correctly, as it turned out) that the work was orthogonal to the axis of interest.

You should take from this chapter the concept of thinking about actions, decisions, or many other motions as vectors in a multi-dimensional space. You can now think of bringing together the actions or opinions of many people, or the actions and interests of many organizations, and producing a useful sum. You may find that some motions or actions are orthogonal to others. Although you do not produce any numbers in this process, and you will usually not write it down, this approach lets you organize your thinking in powerful ways.

7

Brittle Is Bad

After that chapter you may appreciate a rest. I address *resilience* in the present chapter, and you will be relieved to learn that it is the common English word with the common meaning. I contend that the concept of resilience deserves much more attention than it usually gets. In this chapter, we shall proceed from the way the word is used in science to using it in the structure of organizations and in even more general business applications. We shall encounter resilience here in perhaps unusual settings, and I shall exploit the opportunity to hang on some observations about the strength of organizations.

First, let us consider a bar of a metal or alloy. If it is a *ductile* material (like steel) and is heavily stressed, it may gradually deform somewhat, but it retains its strength and can continue to support a load. If it is a *brittle* material (like glass) and heavily stressed, it will break.

Consider a structure like a steel bridge truss that is constructed of ductile elements. Let it also have somewhat more and larger elements than "it really needs" (i.e., more than just meet the loads for which it was designed). If this truss is subjected to unusually heavy loads (a disastrous storm or an illegally heavy vehicle), one part will be stressed too heavily and will deform; but that element will continue to sustain some load locally while the other elements, augmented by those that "were not really needed," join in sustaining the total load. The structure is *resil-*

ient. It has strength in depth, the ability to "bring up reserves" and to continue functioning after a shock or overload.

By contrast, a brittle structure has elements that are not adaptable and has no or inadequate redundancy. When it is subjected to a stress beyond its designed capability, it fails catastrophically. Corporations and other organizations and institutions can be either resilient or brittle. The extreme example of a brittle organization is perhaps a dictatorship, in which no sharing of power has prepared the government for the death or failure of the dictator. Some corporations are brittle because of an unhealthy concentration of decision making in the chief executive officer. And even if decision making is shared, the personality of the chief excutive officer may discourage effective participation by his or her associates and the resilience it produces. Further, excessive downsizing may have cut out reserves of competence.

By contrast, a resilient corporation is managed by people who share goals, information, and experience and can adapt (are ductile) to new stresses and conditions. Further, any downsizing that may have been necessary has left sufficient reserves. Such a corporation can maintain its strength in the face of shocks and stimuli that would destroy a brittle organization. Resilience in no way minimizes the leadership role of the chief executive officer, but it buttresses his role with a resilient structure of associates.

In designing structures such as an automobile body, engineers study systematically all of the conceivable ways in which the structure could fail. They then design sufficient strength to prevent each in the face of the expected loads, environments, and stimuli, with an economically affordable margin of safety. They are designing for success, of course, not for failure, but they identify and consider all of the modes of failure.

Often failure will eventually, after a long life, become so likely that the designer will prepare for failure and will insist that the structure or device *fail gracefully*. For example, the engine controls on a modern automobile are a complicated electronic-mechanical maze designed to accomplish simultaneously acceptable performance, fuel economy, and compliance with pollution regulations. It is highly likely that after scores of thousands of miles this mess will fail; engineers work to provide a "get-home" capability in that event. By failing gracefully, although performance (slower trip to the repair shop), fuel economy, and antipollution all suffer, they do not suffer as much as if a tow truck made a round trip to the breakdown site. And, of course, this graceful failure is much more convenient and comforting for the driver.

A key project in a corporation can fail. It is important that the organization is constructed with sufficient resilience that the project fails gracefully. The corporation may not function exactly as planned, but its integrity and reputation are maintained and its people still respect each other. As MacGeorge Bundy said, "We are all going to need each other again."

Of the many considerations involved in designing a project for *success*, I wish to highlight here only two that build in resilience: First, redundance should be provided. It may be expensive, but it is almost always necessary to buttress the straight-ahead path with alternates and support. Second, the plans should answer all the important "what-ifs," questions like: What do we do if a competitor enters the field with a superior product? What if exchange rates batter our international marketing plan? To leave such considerations until the event means loss of time, which may ruin the project. Fred Brooks has commented on the all-too-frequent poor planning requiring later repair: "There is never time to do it right; there is always time to do it over."

In the early days of the computer, the word "algorithm" was always heard. An algorithm is a set of instructions that are literally executed by the computer; the set can contain thousands of individual operations. In the "batch mode" (which was the only mode known until the early 1960s), a collection of data (perhaps scientific, perhaps financial like bank checks) and an algorithm would be fed into a computer, which then might crank away for several hours. For some large, computationally intense scientific problems, this is still the mode of operation. If a single mistake occurs in the algorithm, the machine halts or begins spewing out nonsense. This is clearly a brittle organization with an ungraceful failure mode and is highly wasteful of both computer time and human time.

Algorithms still govern the internal workings of computers, but computer engineers have emancipated us from being their slaves. Most of us work in an *interactive mode* in which we can see how our wishes are being carried out, correct our mistakes, and devise routes around obstructions. As long as no serious hardware or software failures occur in the inner workings of the machine itself, our application of it and response to our own mistakes on a modern personal computer are highly resilient.

A common example of an algorithm is the array of instructions given to a visitor to a strange city in order to find his or her place of business. The secretary of Mrs. A explains to the secretary of Mr. B: "From the airport you take I-90 west to the third interchange; turn right and go to the fourth traffic light; turn left and go six blocks; turn right for two blocks. You can't miss it." Oh, no? Even if transcription into a letter or communication over a noisy telephone line is perfect, disaster impends. One of the lights may be out or a new one added; the number of blocks may be different if measured on the right side of the road and on the left; and so on. Recovery is possible if the streets are laid out like those in a Midwestern American town, but Mr. B does not have a prayer if the layout is like downtown Boston.

An algorithm is a quintessentially brittle system. The resilient approach is, of course, to give Mr. B a map. Even if he makes a mistake or two, he can find out where he is and recover. (Perhaps the ubiquity of fax machines will make the algorithm scenario obsolete.) In communication and, more importantly, in designing organization structures and business operations, the algorithmic approach is to be carefully avoided. This is especially true in a *business plan;* to the extent that it can look more like a map than like an algorithm, it will be more likely to survive unexpected buffeting. If the plan has addressed all of the "what-ifs" and explored the alternate routes as contingencies develop (as in a map), it will be *robust* and resilient.

This is perhaps an appropriate place to present some notes on the strengths of organizations and to contrast an academic department with a corporation. (Although you, the reader, are not expected to be operating an academic department, you are probably an influential alum or even a trustee of one or more institutions with such.) The strength of each depends, of course, strongly on its people. The academic department will have some (usually young) innovative, imaginative, fearless people who risk their careers by inventing outrageous theories or interpretations of history or literature, or by tackling virtually impossible experiments. Although frequently wrong, they may be right, and by upsetting the applecart of science or scholarship, they are responsible for progress. The department will also have some careful, cautious, experienced people who will patiently conduct extended tests or experiments. Both the academic department and science and scholarship as a whole get their strengths from nourishing both extremes, and all in between. This diversity provides resilience.

It is more difficult to accommodate such a large diversity in a corporation. (It is especially difficult if the corporation has become excessively "lean.") Although there is a wide variety of tasks and required skills, all people must share the basic goals of the corporation. Some diversity enlivens the atmosphere, and it is possible and occasionally necessary to proclaim that "the emperor has no clothes," but it would make a weak and short-lived corporation if the goal of many was to upset the applecart. The trick is to present a cordial reception to innovation and to manage to create and nourish a resilient organization, not a conformist one; there is always the danger of "where everyone thinks alike, no one thinks very much." And, as Harold H. Hall has said: "The future of lemmings lies with the dissidents."

Another difference between the person undertaking a program of science or scholarship in a university and a corporate administrator (or academic administrator) is worth noting here. It is often said about a great scientist or scholar that his or her work was accomplished to such a high standard that it could not be improved. This can never be the case in administration. Even if a key decision was made as well as it was possible to be made, it would always have been possible (if there had been more time) to have used more data and always possible to have talked with more associates and employees, to secure a more enthusiastic reception of the decision. Such additional data and talking may never have visible manifestations, but they enhance the resilient strength of the organization.

While on the subject of the contrasts between "bench science" and administration, I report that I have often been asked what was the most striking difference between being a scientist and being a university president. I had no difficulty with the answer, even though it saddened me to speak of it: When I was doing science, I wanted to know all I could about everything around me and never could know enough. When I became a university president I learned many things, always about people, that I wished I did not know.

This chapter has exported the concept of resilience from bridges and buildings and applied it to organizations. It has outlined the advantages of resilience and the dangers of brittleness. It has explained some of the ways resilience can be built into a corporate structure. It has also provided some only tenuously related comments about the difference between corporation and university managements, which may help you in your role as a trustee of the latter.

8

Spreading Things Out

In making sense of any complex group of objects, we first spread them out and then sort according to some scheme. It may be as simple as the size of oranges: Remember the old joke of the person who was promoted to the job of grading: "I couldn't take it. Small, medium, and large! Decisions, decisions, decisions!" Or it may be the pieces of a large jigsaw puzzle, in which we sort the pieces by edges, sky, water, trees, and so on. The scientist expresses a group sorted according to some parameter as a *spectrum*. In this chapter we exploit the idea and language of spectra as aids to communication, management, and quality control.

The concept of a spectrum is familiar to you as the *visible spectrum*, that part of the total electromagnetic wave spectrum to which our eyes are sensitive. The *general* concept of a spectrum is a varying of properties as a particular parameter varies. For example, in the visible spectrum the color sensation is the property that changes as the wavelength of light varies from about 0.35 to about 0.65 nanometers (1 nm = 10^{-9} meter). A prism or a rainbow spreads out the different wavelengths for us to see and to perceive as different colors, from a deep violet at 0.35 nm to a deep red at 0.65 nm. (It is, of course, no accident that the spectrum of light coming from the sun is almost exactly the same as the spectral sensitivity of the human eye; the magic of evolution has done its work here.)

In other parts of the electromagnetic wave spectrum the eye is useless, but sensitive devices can be built to detect the radiation, which is basically the same but which is given different names in the different regions of the spectrum. Radio waves, microwaves, infrared radiation, visible light, ultraviolet radiation, X-rays, and gamma rays are all electric and magnetic fields propagating at the same speed of light, but they are given different names because in the different regions of the spectrum the means of generation and detection are quite dissimilar.

Another spectrum, and the one we shall exploit more intensely here, is the *acoustic* spectrum of sound waves, vibrations in the air or other medium. The lowest frequency humans can hear is about 20 vibrations per second, which corresponds to a wavelength of about 17 meters. The highest frequency humans can hear is about 17,000 vibrations per second, a wavelength of 0.02 meter. Again, this is only a small part of the total acoustic spectrum, which includes the very low frequencies (long wavelengths) of the rumbling of the earth in earthquakes, the very high frequencies (short wavelengths, or "ultrasound") used to image defects in metal machine parts or to outline organs in the human body, and all frequencies in between those.

We proceed in several different directions from the concept of a spectrum. First, the language serves well to differentiate elements among a family of products, according to size or any other parameter. You can think of a *broad* spectrum or a *narrow* spectrum of products or even of people, distinguished by age, weight, education, or any other parameter. A vocabulary can be thought of as a spectrum; if you use a broad spectrum of words and I know and am sensitive to only a narrow one, communication is confined to the narrow spectrum.

It is perhaps a little forced but nevertheless helpful to think of the possible stimuli to nuclear war as a spectrum. The spectrum so far experienced includes the Berlin Wall, the U-2 incident, the 1962 Cuban missile crisis, Vietnam, the 1973 Mediterranean and oil crisis, the Iranian and Beirut hostages, Afghanistan, Kuwait, and the breakup of the Soviet Union. That spectrum of stimuli has not produced a nuclear war, countered as it was by diplomacy, patience, good sense, and deterrence by the "balance of terror." The trouble is that the spectrum will doubtless grow and become broader. Although the power of the country thought to be our most likely antagonist is dwindling, the cast of characters is becoming more complicated and diffuse. There are now almost a dozen countries with nuclear weapons and scores of countries with ballistic missile technology, capable of delivering nuclear warheads or other weapons of mass destruction (chemical and biological). How does deterrence work, or does it work, in this new situation? Will the counters continue to prevail in the face of a broader spectrum of stimuli?

For my second direction, I return to the acoustic spectrum and introduce the concept of a *filter*. A symphony orchestra generates sounds of frequencies from about 15 to about 20,000 vibrations per second; the spectrum of human hearing is only a little narrower. If such sounds are processed electronically through microphones, amplifiers, and speakers, the resulting sound is reproduced imperfectly. The plots in Figs. 8-1 and 8-2 illustrate the *filtering* actions that can take place. The first plot

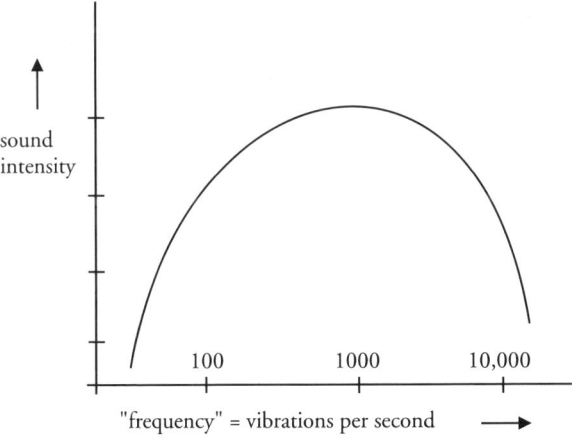

Figure 8-1. Spectrum of sounds from a symphony orchestra.

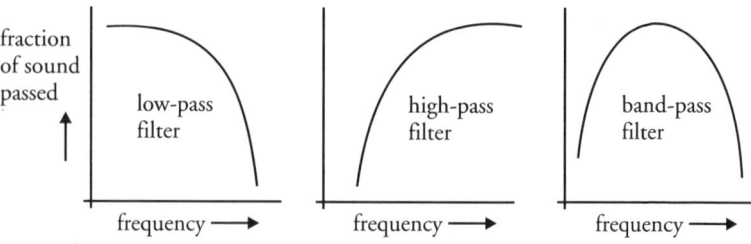

Figure 8-2. Acoustic filters.

shows the spectrum of sounds from a symphony orchestra, averaged over a few minutes.

The next plots (Fig. 8-2) show what happens to these sounds if they pass through the apparatus of recording and reproducing, microphones, amplifiers, perhaps tape or compact-disc recorders, and loudspeakers. In the electronic processing, some frequencies are suppressed; the entire spectrum is not reproduced faithfully. The three kinds of filters are shown: A low-pass filter, as its name implies, allows the lower frequencies to get though but cuts off the higher frequencies. A high-pass filter is just the reverse. The overall system is always a band-pass device. If the band is broader than the source or broader than the ear's response, the music is reproduced accurately ("high fidelity"). But it is very difficult and expensive to accomplish this, and most of us have apparatus that is quite noticeably inferior. Sounds emerge from the speaker only if they are of frequencies "passed" by the apparatus. If the pass band is too narrow, music will be dull and distorted and speech will be unrecognizable.

The human ear is a band-pass device, and the high-frequency cutoff slides down to lower frequencies with age. When this happens, the listener confuses words that are distinguished from each other by the high frequencies in their sounds, words like "death" and "debt." The sensitivity of the ear is shown in Fig. 8-3.

This discussion of filters should serve as a warning to you that every piece of information that passes to you has been filtered, even that coming through your own ears. You must con-

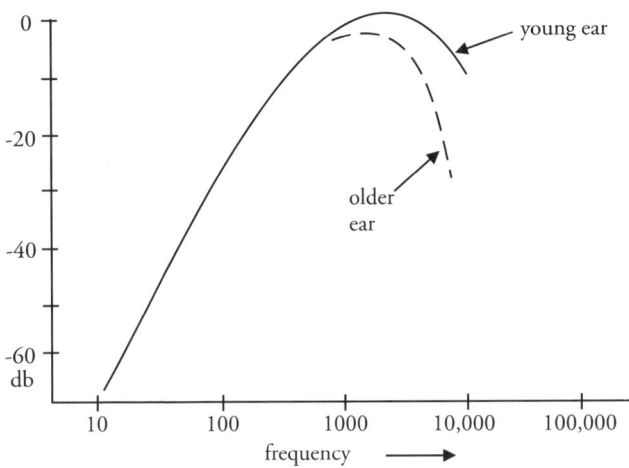

Figure 8-3. Sensitivity of human ear.

stantly be aware of the passbands of the various filters inter-
posed between the sources and your ears and eyes and be alert
to the partial character of the resulting messages. And, of course,
this filtering is far more complicated and extensive than merely
acoustic filtering of speech; you may get much of your information second or third hand through quite imperfect passbands.

The American newspaper is a band-pass device. The spectrum here is broad or narrow according to the sophistication
and prior knowledge possessed by the reader and the sophistication and prior knowledge possessed by the reporter. The supposed narrowness of the former spectrum is often used as an
excuse by the paper to justify the narrowness of the latter.

The restriction of the reporter's spectrum is particularly dis-
tressing to the scientist when it arises from his scientific illiteracy (but I suppose you are at least as upset about reporters'
lack of sophistication in business). Of the many examples of
this disservice to readers, I cite only one, notable not because it
was different from many others but because of the exalted sta-
tus of the reporter. In order to bring some stability into the
rivalry of the superpowers before satellite photography became

effective, the United States employed several high-altitude photoreconnaissance airplanes. The information about missile fields, naval shipyards, and the like thus obtained enabled the United States to respond in patient and measured terms to crises like the Cuban missile crisis of 1962. Without knowledge of military preparations in the closed society of the Soviet Union, the tendency would have been to use "worst-case analysis" which could have been disastrous for both sides.

Earlier in 1962, one of these airplanes, a U-2, was shot down over the Soviet Union. The USSR was extremely proud of the event and exhibited the aircraft wreckage in Gorki Park for the benefit (among others) of American journalists. The head of the Moscow team of *The New York Times* (he later became executive editor of the *Times*) wrote a story under his own byline reporting that the reason the aircraft pieces were so well preserved was that they were made of the "ultralight metal germanium." Now he would not have needed a metallurgist, only a dictionary, to learn that germanium (which he cited several times, it was not a misprint for titanium) is a heavy, brittle, expensive, *non*metal, about the *least* likely element on earth to use to construct airplanes. Did the Soviet propagandists, undoubtedly peeved at the American press's bragging about its freedom, decide that they would find out how much hogwash they could get the scientifically narrow-band filter, the *Times*, to swallow?

When one of the world's finest drama schools was in a complicated uproar, one of the younger faculty chose to leave. Of course the press tried to generate a controversy out of his departure, but the young man resolved that he would not let his departure cause additional suffering. Somehow he managed to get a *Times* reporter to quote him exactly and in full: "The complexities of this situation will not survive the vicissitudes of journalism." For "complexities," read "broad spectrum of concerns and implications"; for "vicissitudes," read "narrow bands." One could perhaps have wished for a less polysyllabic statement, but one must admire the young man both for his self-restraint and his success in upstaging the newspaper.

The limited spectrum of vocabulary creates weird misunderstandings. A passenger on an airport-to-home minibus told the starter that she wanted to go to her home on Charter Oak Street. The starter told her he knew the street well, it was near his own home, and promptly told the driver to go to Chattanooga Street (which was, of course, the street on which the starter lived). We hear only words that are in our own vocabulary unless we make an enormous effort to learn a new one. In effect, we "tune" (adjust the pass band of) our reception to our own limitations.

This circumstance causes a substantial difference between *producers' language* and *consumers' language*. You see examples of producers' language in almost all instruction books (perhaps worst for computers) wherein slang, jargon, labeling of parts by words never heard elsewhere, and even no definition of "front" and "back" are common. You hear other examples as functionaries behind desks or windows (banks, airline check-in counters, auto registration windows, and the like) speak in terms known only to them but claim to be communicating with you. If there is a prize for the most egregious narrow-spectrum stupidity, I think the writers of "explanatory" materials for the Social Security Administration would win it with little contest.

My earliest and perhaps sharpest realization of the difference between these two languages occurred in the lines of students-to-be at the registration preceding a university semester. An official was dividing the line of people *facing* him into two lines, depending on something he read in the materials each person in the queue presented to him. He then said "to your left" while pointing to *his* left, or "to your right" while pointing to *his* right. I guess he was a good hockey coach, his real job at the university, but I wondered how he communicated with the puck knockers.

Have I fallen into the same trap; is this book written in producers' language? I have tried to avoid it, but as I apologized in the Introduction, I have also tried to avoid annoying the reader by repeating what he or she already knows. Reproducing a wide spectrum of sound with high fidelity has become commonplace. Since both music and speech are composed of many frequencies, it is essential that the microphone (or compact-disc reader or phonograph pickup), amplifier, and speaker combination treat all frequencies alike. Distortion enters if the "gain" of the combination (output divided by input) is different for different frequencies.

To study distortion as simply as possible, let us consider as "input" to an amplifier-speaker combination a signal composed of equal amplitudes of two frequencies, one twice the other (i.e., one octave apart). Suppose these are at the low end of the audible spectrum, say 20 and 40 vibrations per second (lowest and next lowest "E's" in human hearing range). Suppose also that the passband of the amplifier-speaker combination is as plotted in Fig. 8-4 (I have plotted only the low-frequency region, since that is all we need); the falling off at low frequencies is unavoidable without large electronic components and great expense, and so this is a likely pattern.

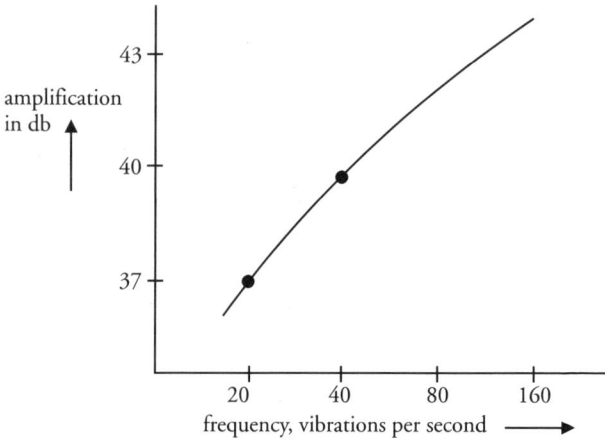

Figure 8-4. An example of an amplifier.

Under these conditions, the higher-frequency signal comes out twice as strong as the lower (recall that 3 db is a factor of 2), and the music is severely distorted. The signal strength is plotted in Fig. 8-5 as a function of time for the input (which is the shape we want) and for the output (which is the different shape that we get with the imperfect system). Our ears easily recognize the imperfection in the rendering of these two tones.

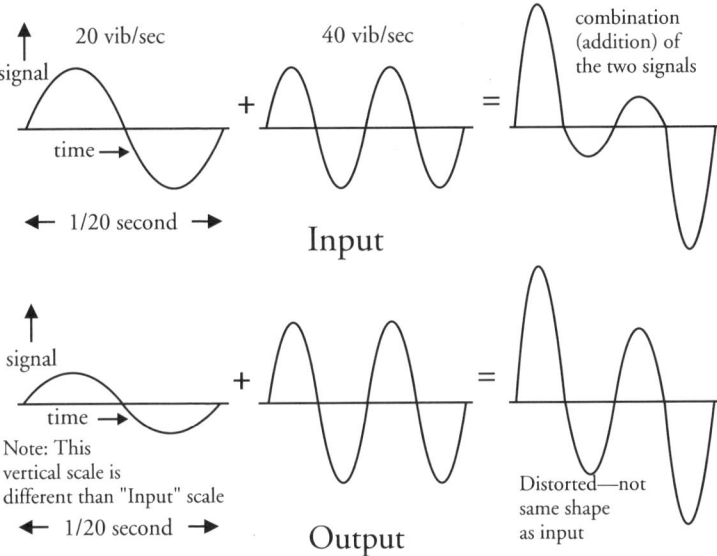

Figure 8-5. The upper curves are the input of two signals to our amplifier. The lower curves are the (distorted) output, with a different vertical scale.

Modern technique has overcome this distortion by the use of *negative feedback*, and this concept of negative feedback is a powerful metaphor that we shall now explore. Feedback is commonly spoken of, as in "getting feedback" from customers about their experience with a product. What we shall describe here is an extension and sharpening of that type of communication and especially of what action is taken in response to it.

The basic idea of negative feedback in electrical and electronic apparatus is to create electrical circuits such that some fraction (much less than one, call it *F*) of the output is compared to the input; the circuits then operate to reduce the difference between the input signal and that fractional output signal to (practically speaking) zero. When this is done, the amplification (you can see that it equals $1/F$) has been accomplished with high fidelity (near zero distortion); the output signal is $1/F$ times the input signal, but the two have exactly the same form and produce exactly the same sound.

Let me try to explain that again in slightly different language. I first note: (1) Amplification is cheap, I can afford to throw most of it away; (2) I want the output signal to have precisely the same shape as the input signal. I set up the electric circuit to give me a fraction *F* of the output (perhaps $1/100$ or $1/1000$). I then make this *F* oppose my input signal, make it subtract from it. I amplify the *difference* many, many times (10,000 or 100,000 times perhaps). Thus the slightest difference between the shape of the input signal and the shape of the output signal is amplified enormously and "fed back" to the input. The process effectively reduces the difference in signal shape (sound) to nearly zero.

In our example, our amplifier, as has already been illustrated, has an amplification of 40 db (that is, a factor of 10,000) for an input signal with a frequency of 40 vibrations per second and 37 db (a factor of 5,000) for an input signal with a frequency of 20 vibrations per second. (You will recall, by way of checking our work here and relating it to Chapter 3, that 3 db is a factor of 2, and since $40 - 37 = 3$, the gain of 40 db, 10,000, is a factor of 2 larger than that of 37 db, 5,000). The relations between the input signal *y* and the output signal *Y* are shown in Fig. 8-6.

To get a typical example, let $F = 1/100$. Then we can calculate the gains at the two frequencies in Fig. 8-7. Note that we now have a gain of only a factor of about 100, but the gain is almost exactly the same at the two frequencies. This contrasts very favorably with the performance of the amplifier without

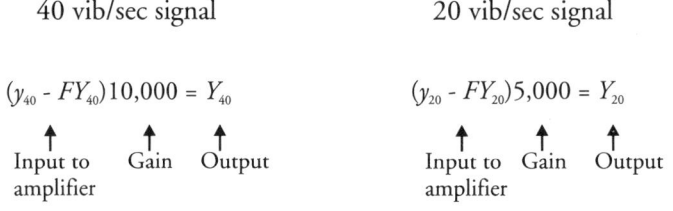

40 vib/sec signal   20 vib/sec signal

$$(y_{40} - FY_{40})10{,}000 = Y_{40}$$ $$(y_{20} - FY_{20})5{,}000 = Y_{20}$$

Input to   Gain   Output   Input to   Gain   Output
amplifier   amplifier

Figure 8-6. Feedback, subtracting a fraction F of the output from the input.

$$10{,}000y_{40} - 100Y_{40} = Y_{40}$$ $$5{,}000y_{20} - 50Y_{20} = Y_{20}$$

$$10{,}000y_{40} = 101Y_{40}$$ $$5{,}000y_{20} = 51Y_{20}$$

$$Y_{40}/y_{40} = 10{,}000/101 = 99$$ $$Y_{20}/y_{20} = 5{,}000/51 = 98$$

Figure 8-7. With $F = 1/100$, the gain is now almost the same at the two frequencies, and therefore there is no distortion.

feedback, wherein there was a factor of 2 difference for the two frequencies and the resulting distortion shown in Fig. 8-5. By making the gain of the amplifier larger and F smaller, we can get as high a gain as we wish with as little distortion as we wish.

The key operation in this negative feedback was comparing the output signal (what is produced) with the input signal (what is intended) and reducing the difference to zero.

I digress at this point to draw another lesson from the history of negative feedback in electronic circuits. This invention became useful only when amplification became cheap, by the invention of clever electronic tubes. Once amplification became cheap, it became possible to throw much of it away in order to have higher-quality performance. Similarly, with the invention of the integrated circuit, computation and memory became cheap (and are getting still cheaper). This circumstance makes possible an enormous variety of "smart" devices. A generalization of this situation is an important lesson: You must constantly be on the lookout for processes and products that are made possible by the major cost reduction produced by some new technology.

Now, back to negative feedback as a metaphor: What we are doing in business is taking the feedback (customer comment, employee concerns, product performance tests, and the like) and *correcting* the process, comparing the output with what was intended and then reducing the difference to nearly zero. In the electrical circuit, the correction is immediate, automatic, and as nearly exact as we care to make it. In human operations, clearly the correction will not be immediate or automatic, and judgment will have to be applied before corrections will be made for all of the feedback signals, some of which may be trivial or wrong.

You will appreciate by now that the scientist's use of feedback is similar to the popular concept, but there is the important difference that the scientist's use emphasizes the step "reducing the difference to zero." It also emphasizes the need to make the feedback loop go around *all* the distorting elements. For a simple example, if you as an executive give instructions that an important communication to shareholders be in their hands by December 15, your feedback should be from a sample of *shareholders*, not from a sample of your communicators.

To give a concrete example, let us consider again the problem posed by producers' language in the example of an instruction booklet for a piece of equipment (a videocassette recorder, a personal computer, an automobile, or the like). The negative feedback approach would dictate a short print run of the booklet and an incentive for the initial customers to communicate instances of producers' language (a word not in the dictionary, an explanation needed at early pages but supplied only at later pages, and so on). Then, with this information, the creators of the booklet would try to put themselves as nearly as possible in the position of the customers (limited spectrum of vocabulary and experience) and rewrite the booklet before great quantities of the product were sold. I emphasize that the feedback comes from a sample of *customers*, not from salesmen or writers, and therefore goes around *all* of the distorting elements.

Quality in the service industry, which is only now getting serious attention, suffers greatly from lack of feedback. When a weekly magazine asks the subscriber to "allow eight weeks" for an address change, what is a subscriber on the move to do, even if the date of the move was known months in advance? I am not aware that any magazine has ever sought to learn what service their subscribers need, although they frequently ask what content they want, what their income is, how much they buy, and other data of interest to their advertisers.

The concept of correction by feedback is, of course, much broader than these simple examples. It becomes most powerful and important when applied to the leadership an executive provides for an organization or institution. Feedback here is marvelously complicated (unlike the simple electrical circuit), with signals coming from a wide spectrum of sources and often internally contradictory. The attitude toward feedback, the skill in obtaining and sorting it, and the wisdom and energy in correcting performance (or at the very least augmenting communication) are essential characteristics in your service as an effective executive. You must tune your eyes and ears to catch appropriate signals from your troops and your customers, just as evolution has tuned your ears to match speech and your eyes to match sunlight.

The concept of spectrum has led us in this chapter to thinking about filters and the damage filtering of communication does in an organization. Although, of course, the electrical circuit does not mimic real life, the idea of comparing input with a fraction of output and feeding back signal to make these two signals essentially identical in form is a powerful idea. You can use feedback effectively to avoid the trap of speaking in producers' language, whereas the only effective communication is in consumers' language. In the next chapter, which is closely related to the present chapter, we will explore further the communication and action process.

9

Detective Work

This chapter applies the scientist's approach to learning about objects, processes, and environments. The approach is through *sensors,* the devices that convert something like temperature into something like an electrical signal, and *channels,* the pathways of processing this information. (The preceding chapter was, in part, a preparation for the consideration of channels.) Both sensors and channels have close counterparts in managing and directing.

The concept of a sensor is already familiar to you. Our five senses detect changes in our environment and monitor our interaction with our environment. Signals are transmitted to our brains and processed there, and then in some as yet mysterious way our "minds" decide what to pay attention to and what to do about it.

Physical sensors parallel closely our human biological sensors. Touch has its counterpart in mechanical probes such as are used by robots to tell where a part being machined is located. Smell and taste have their analogs in chemical detectors such as gas- and liquid-chromatography.

Sound waves are converted by microphones into electrical signals, which can then be carried along wires and through more complicated circuits and processed to suit the need. We looked into this in the preceding chapter for ordinary, humanly audible sound. Other detectors of sound waves operate in regions of the sound spectrum inaccessible to the human ear; very low frequency sound, such as the rumbling of earthquakes, is detected by seismometers; ultrasound (very high frequency sound), for medical imaging and underwater object location, is detected by "transducer" crystals (transducing from sound to electrical signals).

Video cameras and similar devices convert light into electrical signals. The huge width of the electromagnetic wave spectrum (with common usage from wavelengths of 10^{-13} to 10^6 meters) necessitates a wide variety of detectors. The very shortest waves, as I have already noted, are called gamma rays, then waves of length up to about 4×10^{-7} meter are called ultraviolet, waves from about 4×10^{-7} to 7×10^{-7} meter are called the visible light spectrum, longer waves up to about 10^{-3} meter are called infrared, and waves longer than 10^{-3} are called radio waves, with different labels (microwaves, UHF for ultrahigh-frequency, and several others) for different bands of the radio region.

In scientific and technical apparatus, the sensor almost always produces an electrical signal that is the *input* to a channel of electrical processing. In its simplest form, this is just the sound amplifier we have already encountered, but remember that even in this simple situation some filtering always occurred. Much more complicated processing is commonly performed for signals like radar or satellite-relayed telephone signals, usually converting first to digital form and then employing a microprocessor (computer).

Incidentally, it is becoming increasingly common in long-distance communication to convert an electrical signal to a visible light signal and to transmit for considerable distances through an optical fiber, a tiny filament of very pure (and thereby very transparent) glass or quartz. The lower transmission losses (lower attenuation), smaller size, and other advantages more than offset the necessity for the extra steps of converting from an electrical to an optical signal and back again at the end of the line.

The start of the process of learning about objects, processes, and environments is a sensor with adequate *sensitivity*. It must produce enough signal to be worth processing. Its output then enters the channel from a sensor to the production of an instrument reading and a decision (*output*).

The channel always modifies the signal in three ways.

The first way is by *distorting* the signal, as we saw in the preceding chapter; with enough care and cost, this distortion can be reduced to acceptable levels, primarily by the use of negative feedback.

The second way is by introducing *interference* (sometimes called "crosstalk"), which is a signal or signals like the one in question but extraneous to it. The simplest example (which gives it the name crosstalk) is the coupling into our telephone line of signals from other telephone lines in close electrical proximity to it; one rarely experiences this nowadays since the telephone system has achieved high quality, often by using optical fibers for transmission over long distances, but you will still hear it if you tune an AM radio. If you have ever tried to think (or even just navigate) on a city street when someone is exercising a "boom box," a high-powered audio amplifier emitting music (?), you will appreciate the importance of an interfering signal invading your personal space.

The third way in which the channel modifies the signal is by introducing *noise,* and this requires some explanation. All channels exhibit spontaneous fluctuations. In an electrical amplifier, for example, if the gain is turned up high enough, the output voltage with *no* input will look something like Fig. 9-1. A sec-

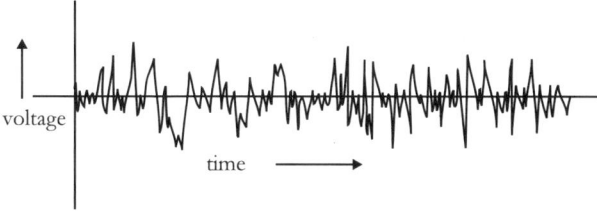

Figure 9-1. Noise.

ond or two later the precise pattern of peaks and troughs will be completely different, but the general character and magnitude will be the same.

If a small pulse of signal is connected at the input, the output will show both the noise and this signal. If noise and signal are of comparable magnitudes, the output voltage may look like Fig. 9-2. Again, this is only an instantaneous snapshot and the details will be completely different a second or two later, but a small signal will still be recognizable.

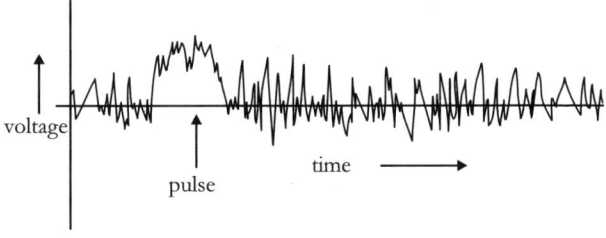

Figure 9-2. Noise plus a small signal.

If this is an acoustic channel, this pulse will be heard as a weak burst and a noisy background. Clearly if the signal is small enough, it will be "lost in the noise," and it will not be possible to tell whether or not there was an input signal. I am not quite ready to apply this discussion as a metaphor, but you should already recognize noise as a distraction in your attempts to receive all sorts of signals in real life. "Lost in the noise" or "down in the noise" are expressions often used by scientists to describe a signal or effect that is insignificantly small with respect to background fluctuations. We shall look more closely at such circumstances when we discuss statistics.

These fluctuations in my amplifier can be larger than in yours if I have a poorly designed circuit or "noisy" components. But there is a fundamental lower limit below which the fluctuation power cannot be reduced. This is an extremely important and absolutely irresistible law of nature that we shall enlarge on somewhat when we deal with conservation laws and with entropy. An often heard expression is the "signal-to-noise ratio" (*S/N*). In the example plotted in Fig. 9-2, the *S/N* is about a factor of 2. For good television reception, a *S/N* of at least a factor of 100 (i.e., 20 db) is required. If the *S/N* is only about 1 or less, the signal is virtually lost in the noise (but if the signal is constant over a long time the *S/N* can be improved enough by averaging the output over that time to detect the signal confidently).

As you have already seen, the label "noise" arose naturally from acoustic channels. A radio or the audio channel of a television receiver when it is *not* tuned to a station provides only noise. The video screen when the television set is tuned between stations also shows fluctuations of light and dark, and those are also called noise (even though seen rather than heard) or "snow."

Let us now put all this together and use the example of a ship's radar. Our ship has a powerful microwave transmitter that illuminates the surroundings. If there is another ship within range, a tiny fraction of this energy is reflected from that ship and a tiny fraction of this is picked up as an input signal into our receiver (Fig. 9-3). We will have built the radar with all the power and capability that time and the budget will permit, but we always want to detect targets at ever greater ranges, which means ever smaller signals coming back to our radar receivers. The "reading," the blip on the screen or a meter reading, will be the product at the bottom of the illustration.

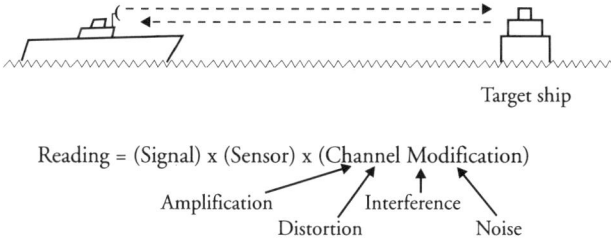

Reading = (Signal) x (Sensor) x (Channel Modification)

Amplification   /   Interference   Noise

Distortion

Figure 9-3. Radar and a signal channel.

In the face of the inevitable channel imperfections, we have to tell whether or not there is a ship out there. We could miss a real ship, or we could conclude that a ship was there when it was not ("false alarm"). It has been reported that at Pearl Harbor on December 7, 1941, the primitive radar actually detected the attacking aircraft but that it had recently had so many false alarms from flocks of birds and the S/N was so poor that the operators were reluctant to send out an alarm. Like the legend of the boy who cried "Wolf!," sufficient false alarms destroy the seriousness with which a real signal is taken.

All of this has been preparation for the metaphor of a communication channel. You can think of a newspaper or of the communication process whereby an executive learns important facts about the operation of his organization as channels. Distortion, the change in the signal caused by the limited passband of knowledge and of sophistication that we discussed in Chapter 8, is probably the most damaging of channel imperfections. Interference, signals that are superficially similar to the meaningful one, can mask what we are trying to learn. And the signal can be "down in the noise" if it is very weak or if the channel introduces abundant noise, with fluctuations in messages constantly arriving at the output (the executive's desk) whether or not there is a significant input.

I have already in the preceding chapter complained of the distortion in American media introduced by the limited passband, especially the limitations imposed by reporters' lack of technical knowledge and competence. I try not to discuss people's motives, but distortion introduced by reporters is particularly damaging if it is a result of the reporter's own agenda. Interference is not so much of a problem in the media, except in so far as the extensive coverage of professional sports and television personalities denies space to more consequential stories. And noise, extraneous signals which make it difficult to extract the real content, is prevalent in almost all articles; a particularly striking example is the reporting of eyewitness accounts at the time of an airplane accident, almost pure noise.

You will certainly have resented the inadequate communication by a pilot on a commercial airliner when speaking to his passengers over the aircraft's public address system. He is presumably making sense to himself, with his earphones on, but he makes no allowance for the distortion, interference (passenger conversations), and noise in the channel. If he thought about the channel degradation, he would enunciate clearly and speak formally, instead of remarking casually in his everyday "slurvian."

The Public Broadcasting System brings to United States television outstanding British dramas, based mainly on the classics of English literature and produced and acted by a highly talented crew. The settings feature most impressive backgrounds of people and sounds. But often the speakers employ a dialect that, although authentically related to the region of the setting, is only tenuously related to Oxonian English. The transmission of this signal through the imperfect electronic channel and low-fi television speaker, competing unfavorably with the background noise, leaves us frustrated with incomprehension. I wonder if the American purchaser of these programs for PBS, who presumably previewed them under much better conditions, ever received any feedback from the poor American viewers.

False alarms constitute the most prevalent traffic of many channels. Consider, for example, the warnings of bombs on airplanes. The Federal Aviation Administration channels develop a warning through their sensors (mostly "intelligence" picked up in public places such as bars and through informers); process it through a channel that adds noise from other intelligence sources; distort it according to how much public fear has been aroused recently; and sort it into regions of the world and airlines. Hundreds of warnings are issued each year. After a bomb incident, there is great public interest in "whether a warning was issued." Clearly you cannot pay any attention to the answer to that question unless you know: (l) How specific was the warning (target, place, time)? (2) How many false alarms of equal specificness occur? These pieces of information are rarely provided by the media.

The ubiquitous metal detectors at the approaches to departure gates of airlines are a common example of sensor and channel problems. If the threshold for setting off the alarm is set low enough, so many false alarms occur that the attendants cannot handle the traffic. If it is set too high, it misses a potentially lethal knife. Whether there is any space at all between too low and too high is (and probably should be) a closely guarded secret. It is rather remarkable that these devices work at all, and they may work more by bluffing the occasional malefactor than by physical detection.

Setting *your* threshold for taking a warning or alarm seriously enough to take administrative action is a difficult but vital task. You cannot afford to miss important warnings, but you cannot have your communication channels cluttered up with many false alarms.

All of this is preparation for a closer look at the communication channels in your company, both "up" and "down." You will certainly have received from many sources advice to keep your up channels open. If you can tolerate one more historical illustration, Julius Caesar's experience when he invaded Britain in 54 B.C. is revealing: Caesar, whose nautical experience had been restricted to the almost tide-free Mediterranean, ordered his boats to be drawn up on the south English beaches just above what he thought was the high-tide line. None of his generals— many of whom knew about the English Channel tides— told him about the monthly tidal rhythm or warned him about the spring (unusually high) tides, and a storm coinciding with a spring tide washed away all his boats.

Walter Lippmann, in *The Good Society*, warns against the exercise of power without adequate "up" communication to take advantage of the competence of the troops: "The head man may get his power from Zeus but does not have Zeus' all-seeing competence, or his power from the people but does not have the sum of their ability."

"Down" communication deserves attention, too. Most people in manufacturing organizations and many in service companies go through life without seeing the output of their work actually being used. If you or your managers can provide exhibits or visits or in other ways show your employees how their effort serves people you will have more satisfied employees. Keeping the channels open to geographically dispersed colleagues like your sales staff presents an especially sensitive situation. The sales force can be easily overloaded with routine announcements, and it is not easy for them to sort out what you really wish them to pay attention to when there is so much noise in the system.

As I mentioned earlier (see Fig. 9-2) a signal can survive a noisy channel if it is constant for a long enough time. If management has a *consistent* message, it will eventually get through to the troops even in competition with massive injections of noise.

To return to up communication, sensors of morale in an organization play a vital role but are a matter of some delicacy. You cannot have spies, but informal talking and listening at all levels and through channels (people) that are as noise and distortion free as possible are essential. Without great skill, however, the signal-to-noise ratio can be low. Recruiting and retention statistics help, but it takes an experienced and profoundly honest "human resources" officer to make a reliable channel. Exit interviews are important since the event of exiting usually arises from a serious weighing of alternatives by the employee and is rare enough that upper levels of management can take the time for it. But in my experience, exit interviews must be carefully checked against other information since the reasons given for leaving are frequently more high-minded and "political" (making a statement on behalf of some employee cause) than strictly accurate.

The application of thinking in terms of sensors and channels is much broader than just to communication. Ideas, inventions, and proposals for new services, products, or organization structures originate at different points and make their way through your company or institution. Think of an idea as a signal, often fragile and small compared to interference and noise, that must be understood and faithfully conveyed through levels of management on its way to the output, your desk. Your attention to the quality and faithfulness of the channel (mostly the competence, honesty, and dedication of the people constituting it) will be handsomely repaid.

Sensors and channels for information and warnings about market and competition vary greatly from industry to industry and even among product lines in the same industry. You must develop your own, and you must be vigilant about their quality and timeliness. There is a terrible tendency to compare "our" product *plans* for *next* year with "their" *current product*, and this leads to smugness, inaction, and perpetual catch-up. (I suspect that this error in timely comparisons was part of the reason for the slowness of the American automobile industry to appreciate the Japanese threat.) Market information is especially susceptible to distortion. An overzealous marketing manager may let his enthusiasm (or ambition) distort data so fundamentally that it is actually dangerous.

Sensors and channels for the appraisal of the quality and relevance of research and development are even trickier than those for markets. There is no substitute for the participation of our research and development people in areas where they compete with other scientists. Although it may seem anomalous to use corporate funds to support work that is then published in the open scientific literature, such publication enables our corporate scientists to be members of the "invisible college" and to share results from other corporations, universities, and publicly supported laboratories throughout the world. It also provides a channel for a sufficiently alert and skillful corporate management to evaluate the quality of our R and D organization and people. (But few in the United States Congress understand how unclassified research leading to publications in the open literature strengthens the country.)

The greatest danger that R and D must protect us against is continuing to make wooden ships or buggy whips. Scientists must save executives from technological surprise, and executives must nourish their channels from their engineers to keep informed of possibly relevant scientific developments. In 1946 the director of research for the Radio Corporation of America gave a speech to representatives of companies that licensed RCA patents in which he said that RCA was the leader (essentially true) in electron tubes ("radio tubes," "vacuum tubes") and that "nothing can replace the electron tube." The scientists in his laboratory, like many scientists elsewhere (in the "invisible college"), *already knew* (from progress in basic research on semiconductors) that solid-state devices would be invented in more than one laboratory in a matter of just a few months, and that within a few years the electron tube (except in a few specialized manifestations like picture tubes) would be completely obsolete. The most optimistic interpretation of his remarks is that he was exaggerating for business reasons; the more likely, and very frightening, interpretation is that he was sincere and did not know.

The concepts or metaphors of sensitivity, distortion, interference, and noise are useful in the decision-making process, especially if meetings of staff or managers are involved. Each person's communication of facts should pass scrutiny with these considerations in mind. Moreover, the decision process itself should proceed with high sensitivity, low distortion, and minimal interference and noise.

At some risk that you may be easily scared, I now add a footnote to this chapter to introduce two words that may be new to you: "convolution" and "convolved." They are healthy English words but are rarely used (outside of mathematics), and are used to mean something twisted; a related word, "convoluted," is more frequently encountered, but usually in a pejorative sense. The mathematical process of convolution, which you need not understand in detail, is the process of taking some signal, multiplying it piece by piece by another function, and then adding up the products over an interval of some space. (You are familiar with a "weighted average," which is a convolution of measurements with weightings.) Remember that space can have many dimensions and meanings—it need not be physical space. It can be time or (in our use) the many dimensions of human interactions and corporate management and planning.

In this language, what we have been doing in this chapter is convolving our signal (information from sensors) with the distortion, interference, and noise functions in a communication channel and summing over all the ramifications of human performance. The concept of a convolution is a powerful mode of thought that deserves wider currency.

You should appreciate from this chapter how a scientist thinks of sensors and channels. Channels of communication are the most obvious sort, but channels can also be channels of action. The imperfections in channels (distortion, interference, and noise) must be sorted out and understood by you, the executive. Noise, which is always present, can mask communications and create false alarms. A channel that emits extensive false alarms is worse than useless. My hope is that after reading this you will give more careful attention to your own specific channels and to setting the threshold above which people in your organization call out a proposal, an opportunity, or a problem to your level and expect a response.

10

"Everything Is Relative"

Overheard at a cocktail party: "I know all about relativity: Everything is relative." In the Introduction I promised not to indulge in such chatter, but I need this piece of it to open this chapter. The speaker, of course, knows nothing about Einsteinian relativity, a marvelously intricate and difficult science essential to cosmology and to the remote, semimythical world of the particle physicists. This world is full of red shifts and twin paradoxes, and you will hear a lot of it (mostly wrong) if you drink enough at parties. But we shall have none of that *real* relativity here; it is beautiful but too abstract for our digestion and has almost no connection with our (the real, touchable) world.

The relativity we shall exploit is *Newtonian relativity* and is very close to this chapter's title and its opening caricature. The basic idea is simplicity itself: To specify any position or analyze and predict any motion, you must first specify a *frame of reference*. For example, a set of coordinates fixed on the surface of the earth with the zero at a particular place serves well as a frame for terrestrial navigation. In this "Newtonian" frame, Newton's Laws of Motion hold almost exactly ("almost" because of the spin of the earth, which requires corrections too small for us to worry about here but which we shall find interesting in Chapter 16). For example, Newton's First Law of Motion states that a body continues in steady motion in a straight line with constant velocity unless it is acted upon by an external force.

If we are so brash as to make our frame of reference the deck of a pitching and rolling ship or aircraft, Newton's Laws do not appear to us to hold. But if we add to the *apparent* motion that we observe the motion of our frame relative to a Newtonian frame (the earth), all is well. If we are clear and careful about our local frame and correct all observations by including the acceleration of this frame with respect to a Newtonian frame, Newton's Laws hold nicely, and we can understand and predict motion.

A revolving platform like a merry-go-round is a distinctly *non*-Newtonian frame of reference, and when on one we get the impression that there is a new force ("centrifugal force") and that Newton's First Law is violated. In fact, of course, our impression is just the fault of our unfortunate choice of reference frame. We can watch closely a marble on the slippery floor of a merry-go-round, with no forces on it other than gravity and the support by the floor, both of which are at right angles (orthogonal) to its path on the floor and do not affect its motion. We then observe that it moves in a straight line in the (Newtonian) frame of reference fixed to the earth.

The metaphoric application of this little essay on dynamics is the lesson that the *choice* of reference frame is vitally important. It follows also that one should not change reference frames casually, only explicitly and for good reasons. An important example of the importance of the choice of a frame of reference is the assessment of the performance of a mutual investment fund, a pension fund, or an endowment fund managed by a firm of investment managers. Some of these managers tend to change their reference frames *after* a year's performance and change them artfully to make their activity look good. This practice is easy since there are many frames to choose from and each has much to recommend it; even just for domestic common stocks there are the Standard and Poors 500, the Dow Jones Industrial Average, the Frank Russell Universe, the Hambrecht and Quist Index, the Lipper Index, Morningstar, and many others. Clients must be extremely vigilant and hard-nosed to insist on choosing the frame *in advance* and keeping it constant.

The choices of the starting point and direction of measurement can be exercised profitably in many corporate contexts. For example, the choice of a base year against which to compare profits or productivity can be chosen to illuminate the effect of management decisions. Again, this should be kept constant for several years and then changed only for good reasons.

The choice of reference frame is obviously important in comparing our corporate performance or engineering achievements with that of others. For example, if we manufacture starter motors for aircraft engines, should we choose as our peers the category "electrical equipment" or "aerospace?" If we choose the "industry" we are in, do we choose the domestic or the worldwide industry? If General Motors chose the domestic frame of reference (Ford and Chrysler), it is easy to see how they missed what was happening in Japan.

The proper, advance choice of reference frame is also important in assessing personnel performance or the performance of a corporate unit. Selection of the appropriate time interval is also vital, and I have already noted in the preceding chapter the danger of comparing "our" product or research in one period with "theirs" in an earlier one. Too often analysis of the optimum frame of reference and time interval is attended to only after it is too late to collect the data. (When we discuss statistics in Chapter 12, we shall visit a grimly amusing example of the choice of frame in executive salary comparisons.)

The *direction* of measurement in our reference frame is also consequential. As a trivial example, consider a large brandy snifter. If it contains an ounce of $100 per liter cognac, it is natural and appropriate to think of measuring up from the bottom. If it contains water that you wish to carry without spilling, it is natural and appropriate to measure down from the rim.

In measuring corporate performance quantities like sales, earnings, return on investment, or productivity, it makes sense and may be legally required to measure (and report) these amounts *up* from where we are now. But for management purposes, it is far better to measure *down* from our plans and aspirations. Then, if we get close to our aspirations, we can raise them. As Browning wrote:

> Ah, but a man's reach should exceed his grasp,
> Or what's a heaven for?

Scientists frequently make progress by approaching a problem and measuring in a different direction, or even turning a problem upside down. You already appreciate the profound difference in viewing the activity of an organization from the top down as compared to looking at it from the bottom up. I do not know what to make of the well-publicized cases where an executive works for a period at the bottom of a corporation. Are they serious attempts to learn experience that cannot be obtained in other ways? Or are they stunts? I believe that most, perhaps all, of the benefit of such role reversals can be obtained more readily in other ways, such as visits to the shop floor and observation of workers' use of their spare time. If you have adult children, listening to their friends can provide a "bottom-up" view.

You can say that your repair of a road consists of moving a hole from the center of the road to the side of the road. That example of a reversal of the usual language is perhaps only cute and unproductive, but you should be warned that the entire solid-state device industry of computers and communications rests on "holes," the *absence* of electrons, in semiconducting crystals. Recall that Sherlock Holmes, in "Silver Blaze," spoke of

> "the curious incident of the dog barking in the night-time."
> "The dog did nothing in the night-time."
> "That was the curious incident."

The upside-down or offbeat thinking or concentrating on the absence of something that should be present are often ways of providing a fresh view and stimulating imagination or invention. On a less serious note, in addition to the usual coffee break at management meetings, your personal plumbing benefits from a negative coffee break.

To return to the importance of the choice of the point relative to which we measure, the manufacture of precision parts provides a revealing example. Consider a turbine blade for a jet aircraft engine; it is a complex shape of an expensive alloy. As it goes through a set of manufacturing operations it is carried by a "jig" (or "fixture"). The complex shape is firmly fixed into the jig, which has plane sides and right-angle corners. The jig also has several *fiducial holes,* small holes accurately placed with respect to the blade. At each machine, pins that are fixed in the reference frame of the machine lock into these holes, and the machining operations are then performed with accurate distances and directions from the fiducial holes. The closer a particular cutting or shaping operation is to such a hole, the less accurate the determination of distance can be; if it is only 1 millimeter away, even if the precision is only 1%, the accuracy still holds to 0.01 mm.

A second, more familiar example is the leveling of an elevator when it comes to an accessed floor. This is readily controlled to a millimeter by controls that are responsive to the separation between fiducial points on the car and fiducial points at the floor.

In human affairs, *principles* serve as fiducial marks. The nearer your actions can stay to principles, the more confident you can be that your actions will meet with enduring respect, by you as well as by those you serve.

Principles played a vital role for academic administrators during what Allen Wallis has called "the great campus craze" of the 1960s. (I called it, perhaps overgenerously, the period of "student shenanigans.") Campus unrest was partially stimulated by the Vietnam war, a misguided adventure in support of a non-government in a country where the French had successfully prevented the creation of a central government. Certainly the draft was also a factor. But since the greatest student violence was in Paris, one is entitled to be skeptical about the total array of causes.

One principle that served administrators and faculties well was: Respond to ideas, data, and even opinions (with appropriate sorting and discounting), but never respond to pressure. Responding to pressure stimulated more pressure and rewarded the worst element in the protesting groups. The agony and tragedy at many universities, leaving scars for a generation, arose primarily from failure to adhere to this principle. This principle deserves more attention in corporate life, too. Near the top of a corporation the pressure can become immense, and you would do well to follow this principle.

The Maine lobsterman is enormously and justifiably proud of his local knowledge and his ability to navigate in the complicated and dangerous waters of the Maine coast. Rock formations, which he calls "ledges," abound, and some that can destroy a boat at half-tide or low-tide are invisible at high tide. If he comes to grief on one of these in the fog, his pride is engaged, and his excuse is: "The ledges move around in the fog."

In the 1960s fog of politicized relations with young people (and some who were not so young but trying to get their youth again), almost everything moved. It was therefore especially necessary to stay close to principles, the only lighthouses (fiducial points, reference frames) one could trust.

Yes, everything is relative. But the principled person's behavior operates only in the secure territory near fiducial fixed points. You will doubtless have understood this long ago, and therefore this chapter has been of limited benefit to you. But we have also explored here the choosing of reference frames and points of departure for measurement, which applies to a wide range of corporate and organization problems.

11

It's a Chancy World

This chapter will be not so much metaphor as a little teaching and a lot of warning. In common with the other chapters, however, it explores the contribution science and elementary mathematics can make to thinking, understanding, and decision making. In the language I have been marketing to you, this chapter provides some of the projections of the elementary mathematics of *probability* on the directions of real-life thinking and deciding, projections that are strongly positive and rich indeed.

It is one of the disasters of the American educational system that most students can graduate from high school, from college, and even earn a doctoral degree in the humanities or a law degree without ever learning the most elementary concepts of probability. Since most of the great issues of modern life such as environmental protection, disease, nuclear war, and safety questions of all kinds are understandable only in probabilistic terms, the citizenry is ill equipped to participate in the study and resolution of these issues. And although we may not like it, probability enters substantially into business decisions and operations.

The mathematical theory of probability originated from card games, and I am counting on your understanding of that point of departure. The simplest example is the canonical five-card draw poker: If I have been dealt five cards including four hearts, what are my chances of discarding the card that is not a heart and drawing a heart? Answer: There are 47 other cards, in the pack and in other players' hands; 9 of these are hearts. I have no reason to believe (I have not peeked) that a heart is more likely to be in one place or the other. My chances of drawing a heart (filling the "flush") are 9 successes out of 47 tries. We say that the *probability* of drawing a heart is 9/47, which is about 0.19. If I carried out this procedure many, many times I would expect that I would succeed in approximately 19 out of every 100 tries.

Probability thus is a number always lying between 0 (certainly not, no chance) and 1 (certainly yes, guaranteed), which gives the chance of "success."

The probability could be confidently calculated in this example since it was easy to identify *equally likely cases*. As we shift our attention from card games to more complex problems, it becomes harder and harder to identify equally likely cases, and thereby much of the confusion and controversy of social questions involving probability arises. Before moving to this area, however, we shall explore calculations of probabilities a little more.

I can introduce the calculation of the probability of a *succession* of events through the example of tossing a coin. If it is an honest ("unbiased") coin, the probability of heads on the first toss is 1/2. What is the probability of heads on both of two tosses? Answer: The first toss can be either a head or a tail; the second toss can be either a head or a tail; there are thus 4 equally likely outcomes of which one is a success. The probability is therefore 1/4.

There is another way of looking at this and that is to say that a probability of heads the first time is 1/2; subsequent to that toss, we toss again with a probability of 1/2; the probability of the combination (both heads) is therefore $1/2 \times 1/2 = 1/4$. In calculating the compound probability by multiplying the individual parts, we have tacitly assumed that the two parts were *independent*, that is, what happened on the second draw did not depend on the outcome of the first. We shall next explore independence with some care.

What is the chance of drawing an ace in a single draw from an honest full deck of cards? The probability is 4/52. What is the probability of drawing an ace from one full deck and an ace from another full deck? The outcome of the second draw is clearly independent of the outcome of the first. The probability is 4/52 × 4/52.

But what is the probability of drawing two aces in succession from the *same* deck? (We do not put the ace that has been drawn back into the pack.) The probability of an ace on the first draw is still 4/52, but the probability of an ace on the second draw is now 3/51, and so the answer to our question is 4/52 × 3/51. We can check this easily by direct enumeration of the equally likely outcomes: There are 4 possible cards that we could draw on the first draw that would lead to success; there are 3 on the second. The total number of ways (ace of clubs first, ace of diamonds second; ace of clubs first, ace of hearts second; and so on) of succeeding is 4 × 3; similarly the total number of possible outcomes (cards that could be drawn) is 52 × 51. Thus by direct calculation of the ratio of successes to total outcomes we reach the same probability, (4 × 3) ÷ (52 × 51). Here the second draw was *not* independent of the first.

There is a curious misapprehension about independence in a popular interpretation of the "law of averages." Many believe that if an unlikely run of drawing cards or tossing coins occurs, the next draw or toss will somehow take account of the history and move the series closer to the expected probability prediction. I knew a Ph.D. who later became president of a major university who seriously believed that if a tossed coin came up heads 50 times in a row, it was more likely to come up tails on the 51st throw, that the law of averages compelled that conclusion. There is, of course, no way the coin can "remember" the results of the first 50 throws; they are completely independent and the succeeding throws are independent of them.

If this coin-tossing game continued for 500 or 5000 throws, one would expect results like (among many, many others) 270 heads, 230 tails, and 2460 heads, 2540 tails. The law of averages works by *overwhelming* any unlikely run of results by succeeding, more likely outcomes. The larger the number of throws, the closer the fraction of heads will come to 0.50, but the absolute difference between the numbers of heads and the numbers of tails will, on the average, continue to grow. (Of course, if some clever metallurgist had milled and lapped the heads-half of each of two coins and sweated them together, the 51st throw would be certain to be a head!)

I hardly need warn you of the business implications of this: If your business has had (you believe) "a run of bad luck," you must not rely on the law of averages to assure better performance in the future. If anything, what looks like a run of bad luck is a result of a fundamental weakness (like the coin doctored by the metallurgist) that must be cured.

I next dispose of a jocular case of nonindependence in which the compound probability should not have been calculated at all. This is the joke (?) about the bomb on the airplane: A passenger was told that the probability of a bomb's being on an airplane was one in one million, 0.000001. He asked: "What is the probability that there are two bombs on an airplane?" and was told that it was 0.000001×0.000001, so now he carries his own bomb with him! The silliness here is obvious, but it is not always obvious when some knowledge must be substituted for a probabilistic calculation; in this case, 1 (certainty) should, of course, have been substituted for the 0.000001 for the bomb he carried.

A less obvious fallacy occurred in a three-engine aircraft on a flight from Miami to Nassau. When very near Nassau, one engine's instruments showed falling oil pressure, and the captain naturally shut down the engine. He turned around to return to Miami, since that was his maintenance base and since his procedures book told him that the probability of a required inflight shutdown of an engine was only about 1/10,000 per hour, so small as not to create a substantial risk that another engine would go out on the short flight back to Miami. But that calculation assumed that a second shutdown would be *independent* of the first. In fact, all three engines went out and the plane barely and only luckily landed safely at Miami. It turned out that the performances of the three engines, far from being independent, were intimately related in that a maintenance mechanic had made the same mistake on all three engines (he left out the O-rings that sealed the bottoms of the oil reservoirs). (There were complications involving Nassau weather that may have partially absolved the pilot.)

At a golf tournament in 1989, four holes-in-one occurred on the same hole in two hours of play. The media found somewhere a probability of a hole-in-one; I do not know how they did this, but I am afraid they may have divided the total number observed by the total number of holes played in some unspecified area and over some unspecified period of time, which, of course, would give no effect to skill of players and difficulty of the holes. They then multiplied this small number by itself four times to get a preposterously low "probability" of the four holes-in-one! Of course, the event could not in any way be analyzed in probability terms: It was a short hole; all the players were highly skilled; the pin placement on the green was strongly favorable; it had been raining heavily for days, and the green exhibited many of the characteristics of a well-lubricated bowl with the hole at the bottom.

To avoid such foolishness you should always scrutinize the underlying information on which an alleged probability is based. Probability can be *quantitatively* specified *only* if you can examine an array of *equally likely cases*. In the playing-card games with an honest deck and dealer, drawing each card is as likely as any other. The farther you get from this safe territory the more skeptical you must be. You must factor in any knowledge there is and not try to substitute probabilities. You must be alert in multiplying probabilities of successive events and verify that they are, in fact, independent.

But unfortunately in the commonest and most consequential

uses of probability (e.g., the chance of rain or the likelihood of a particular accident), we are far removed (a long distance in this decision space) from equally likely cases. We cannot forgo probabilities, but we must tread carefully and put wide "error bands" on any probability we assert (e.g., we might say that the probability of a particular event lies between 0.01 and 0.001 per year).

Probability implications for business and for most other affairs include not only the chance of some happening, but in addition the consequences if it does happen. The simplest form of this occurs when the consequences can be expressed quantitatively, for example as a number of dollars. The *mathematical expectation* is the product of the probability that an event will occur (a success) and the payoff if it does occur.

To introduce this concept with as simple an example as possible, consider the following game: You are to name the number of a playing card (not the suit) and then draw a card from a full deck. If you draw a card with the number you chose, you receive $10. How much should you pay for each draw if you wish to break even on the average? Answer:

Probability of success = 4/52 = 0.077;
yield if successful = $10;
expectation = 0.077 × $10 = $0.77.

If you pay $0.77 for each of a large number of draws, you will approximately break even.

For less idealized situations in, say, business or national defense, the probability of success is usually the product of several ingredient probabilities. That is, several individual events must occur in order for there to be a success. For example, in estimating the expectation for a new product, *at least* the following enter: the probability of a successful development; the probability of an effective marketing staff; the probability of the absence of a prior competitor; and the probability of the absence of some other product that makes ours obsolete. Each of these probabilities is, of course, less than one. Even if each of four probabilities is, for example, what appears to be a very healthy 0.8, the probability of success is only 0.41.

The most common error is likely to be *not* misestimation of one of these factors, but the failure to identify and include some additional factor, which is always less than one and may be substantially less than one. If our company's disappointment in less-than-expected rewards from new products seems to be more prevalent than unexpected successes, the reason is probably to be found in the frequent omission of relevant factors in the product of these probabilities.

The situation is similar when estimating the probability of success of an acquisition. A particularly costly example of failing to put enough probability factors into the estimation occurred when a large and otherwise successful and well-managed company acquired a semiconductor company. The probability that the industry would quickly transit from the high margins characteristic of an industry introducing new technology to the low margins and punishing competition characteristic of a commodity industry was not included. The ink was hardly dry on the acquisition papers when the transition occurred.

An interesting and famous illustration of this possible error is Pascal's Wager (also known as Bacon's Infinite Conjecture). In the seventeenth century, Blaise Pascal "reasoned" that the probability of entering heaven was proportional to the amount of good works and exemplary behavior one practiced on earth. The reward if one was admitted to paradise was clearly infinite. He thereby viewed the mathematical expectation as a finite probability multiplying an infinite outcome, which gives an infinite reward, and therefore one should spend his or her life doing good works. Since the official word at his time was that the probability that there *was* a paradise was large, maybe even unity (certainty), he did not multiply by this probability. Somehow, ordinary people sensed a fallacy: The widespread lack of good works and exemplary behavior has ever since revealed the discounting by many of the official view.

Mathematical expectation has other tricky elements that you should consider: Is the game long enough? Is the "bank" large enough? The eloquent illustration of this caution is the Petersburg Paradox. This is a coin-flipping game in which you continue to toss a coin until a head turns up, at which point the game ends. If a head turns up on the first toss (probability of this = 1/2), you receive $2. If on the first toss the coin is a tail and on the second a head (probability = 1/4), then you receive $4. If there are two tails and then a head (probability = 1/8), you receive $8, and so on. How much should you be willing to pay to enter this game? Your expectation is: $(1/2)2 + $(1/4)4 + $(1/8)8 + . . . = $1 + $1 + $1 + $1 + . . . = $infinity. Thus it would be a bargain to pay *any* finite amount to enter this game! The paradox lies in our intuitive feeling that paying more than a few dollars would be a big mistake.

The resolution of the paradox is that, yes, the expectation is infinite, but only if you are patient enough to carry out the game an infinite number of times and also, of course, if the resources of the bank are infinite. Even to get a return of only $100 on a single game requires, on the average, playing the game about 50 times. Suppose you had paid $100 a game to enter; you would have had many, many losing games before you had a game with a substantial reward. And only if the bank is extremely strong do you get paid for the (extremely rare) toss that redeems all those miserable $2 and $4 tosses.

The Petersburg game has no exact counterpart in real life, but it cautions us in business and personal finance decisions not to put too heavy reliance on large payoffs that will occur only under conditions that, if we estimate probabilities realistically, are quite unlikely. Also, in deciding whether to invest in the development of a new product, we must carefully consider the size of the ultimate targeted consumer base, the analog of the bank above.

This Petersburg situation is a common occurrence with new programs or new products in industry. The promoter of a new product or program may predict that his game will go on for a long time, with steadily increasing rewards, because the initial development and product introduction costs will be recovered early and thereafter the profit per unit will be large. But realistically it is more reasonable to predict that the Petersburg game of large rewards will be cut short by competition or obsolescence, and the net gain of the whole program may be negative.

The U.S. Congress, of all organizations, is perhaps the least able to live with uncertainty. A congressional hearing will demand that a witness (generally called "Doctor"!) say that some accident or other disaster is "impossible," not just "unlikely with a probability of less than 0.001 per year." Real scientists and other responsible expert witnesses are reluctant to use such absolute terms, since few accidents, diseases, or other disasters are actually impossible. This congressional behavior may introduce an element of healthy conservatism into environmental, medical, and national security affairs, but it can also be very wasteful of the nation's resources. Also, I am reluctant to criticize the congressperson, since he must reassure his constituents, almost all of them innocent of the most elementary concept of probability, that he can *guarantee* their safety, not just enhance it.

The search for certainty convolved with scientific illiteracy to produce a major consequence at the Yalta Conference in February of 1945. Roosevelt and Churchill bargained away territorial and other "chips" to secure Stalin's participation in the Pacific war. But according to Hans Bethe and others, the Manhattan Project had already proceeded far enough that the Los Alamos scientists knew that the probability that the first test of an atomic bomb, scheduled for the summer of 1945, would be successful was at least 0.9, and that this probability would approach 1 (certainty) after another month or two of development. The political leaders discounted the probability, presumably because they were accustomed to hearing promises that in the event were not fulfilled.

In addition, the scientists knew that the yield of an atomic bomb would be absolutely unprecedented. Again the political figures did not credit this prediction, which an undergraduate physics student could easily understand and believe, again presumably because they were used to the exaggeration by promoters of programs, including the exaggeration by scientists promoting science programs.

Both of the two terms in the mathematical expectation (the probability of success and the consequences of success) were thus drastically underestimated, and the political leaders gave away more than they needed to give, with long enduring implications for eastern Europe.

In July of 1979, the Federal Aviation Agency closed the Washington National Airport for an hour (no planes were allowed to land) because of the fear that debris from the satellite Skylab, which was about to disintegrate, might hit an aircraft. Now the probability of such an accident is very difficult to calculate, since it is a combination of known facts about the orbit of the satellite and unknown facts about the precise interaction with the upper atmosphere when it begins to reenter the atmosphere. But even a quick estimate shows that the probability of damage to an aircraft near Washington was so vanishingly small as to make calculation unnecessary. And, again, there was also a foolish error in determining the consequences if Skylab debris showered on the Washington region. Somehow the FAA assumed that allowing aircraft to land and to be on the ground was more dangerous than to have them hit while flying around in a holding pattern. (Skylab actually came down in the Australian desert.)

Finally, I wish to remind you of a rarely heard English word, "stochastic." Stochastic means having an element of chance. There is a stochastic element in most things in life, and we must learn how to live with it without becoming paranoid.

We usually think of the stochastic features of life as all bad, but that really is not correct. The stochastic element in card games, for example, provides their social acceptability: Without it, losing would make them as unpopular as chess. At least at the present state of knowledge of human biology and of detailed knowledge of our human interiors, dying has a strong stochastic component. And this is probably also good, since it would be an ugly world if each of us knew the time of his or her own death.

There are business situations in which it is desirable to introduce artificially a stochastic element. One of these is the internal audit of a corporation. In addition to auditing parts of the organization on a regular, recurring basis, it is worthwhile to audit a few units occasionally at times chosen by chance, to make sure that those tempted by fraud or sloppiness cannot count on an uninterrupted period without an audit.

But in business the more frequent interventions of stochastic elements are changes in the environment affecting our plans. There is always a stochastic element in the playing out of the plans for a product or an acquisition. It is this circumstance that has led to the oft-repeated dictum, popularized by General Dwight Eisenhower: "I have confidence in planning but not in plans." If the planning process is done well and if the assumptions of the environment (inflation rate, foreign exchange rates, etc.) are made explicitly, revisions can be made in response to the buffeting of chance. The assumptions will never be exactly fulfilled; each time they change, the new circumstances can be put into the plans, and they can be revised in timely fashion.

As I explained at the beginning, this chapter is more warning than "how to do it." I have warned you about numerical calculations of probabilities when the input information was a long way from card games, about independence of successive events, and about the necessity for making sure *all* the probabilities are included in business estimates. But nevertheless you must live with probabilities since they are increasingly important in modern life and business, and I encourage you to refine and extend your thinking in probabilistic terms.

12

Statistics: Extracting Meaning in a Complex Environment

The use of statistics, both in description and in statistical inference, is indispensable to extract meaning from activities in today's environment. In business you must use statistics in many different ways. In this chapter we explore some of these, and I warn you about some of the most important traps that others are laying for you. Statistics and statisticians have a bad name, and yet the modern world is dependent on and enormously benefited by mathematically sound and intellectually honest statistical analysis. The bad name arises because there are multitudes of ways for the careless, the self-deluded, or the dishonest to misuse statistics, to mislead the public, and to "prove" relations that are simply wrong. Unfortunately, charlatans can "prove" anything they wish by the improper use of statistics.

In this chapter, we shall examine first the role of statistics to *describe* events, populations, or processes. This is relatively safe territory. We shall then move on to the more treacherous area of *sampling* and finally to the really dangerous ground of *cause and effect*.

I shall begin with some comments on the use of statistics solely for *descriptive* purposes. Both the pitfalls and the power of statistics are less in this mode. Then I shall explore just a little of the use of statistics in process control and statistical inference. Here is where statistics have become indispensable but where the field is littered with booby traps and land mines. I cannot warn you against all, or even many, of these dangers, but I hope to enlarge your appreciation of the language and utility of statistics.

The scientist is trained to be skeptical about every step in taking and analyzing data. You as an executive share this skepticism when you draw conclusions from reports that are far removed from the raw data.

Were the individual measurements accurate? The measured quality of the product you are manufacturing depends on many individual measurements, and measuring instruments must be checked and calibrated; even the accuracy of sales figures is questionable if the opportunity for returning products reduces a "sale" to a mere consignment. How was the selection of data made? Yearly or monthly aggregates are usually safe enough provided that no data were discarded, but aggregation in other ways can be a tricky process, especially if any sampling played a role. Are there enough data? Before drawing any conclusions from averages or other numbers derived from the data, the sufficiency of the data must be demonstrated.

This skepticism is especially necessary if the handling of the data might be contaminated by the wishes of the handler. I once was subjected to an oral presentation about an "experiment" that began: "The purpose of this experiment is to show that . . . ." Of course, the person talking, although enjoying an impressive title and labeled by impressive degrees, was revealing that he did not know what an experiment was and that he certainly was not a real scientist. Because he made clear how he wanted the "experiment" to come out, it was impossible to attach any significance to his work.

Any demonstration ("show that . . .") must be sharply contrasted with an experiment, in which it is absolutely necessary to start with an open mind; contamination by thinking (or wishing) that the data will come out in a particular way is fatal in a real experiment.

But let us suppose that we have good data. For an introductory example I have manufactured some data on the heights of the men in a certain group of 1000, which we call the "population." I plot in Fig. 12-1 the number in each interval of height (say, each 1 inch) vs. the height *h*. (These are not real data of any real population; note also that I have suppressed the zero of the horizontal scale in order to see the detail.) We call this a *distribution*; in this case it is the distribution of the heights of the men in this population. Only one point is shown, but each point is the number (on the vertical scale) of men who have heights within 0.5 inch of the height number on the horizontal scale. Then the curve is drawn through the points. The point shown plots the fact that 70 men have heights between 65.5 and 66.5 inches.

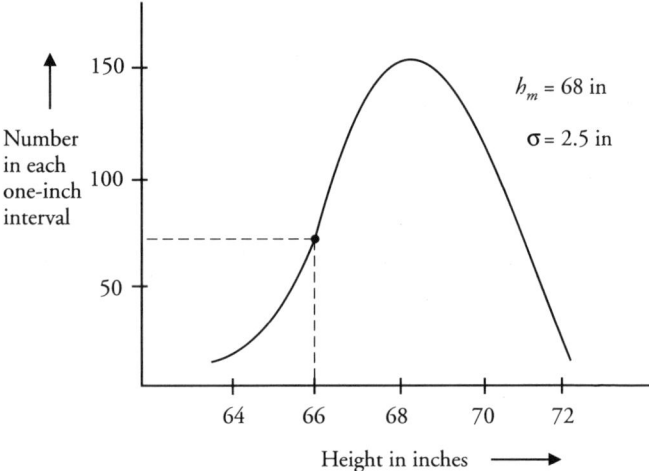

Figure 12-1. Distribution of the heights of a population of men.

There are two statistical measures of this distribution that are common and useful. The first of these is the *average* or *mean*, which I will call h_m, simply the sum of all the heights divided by the number (1000) in the population and a measure of the position of the distribution along the horizontal scale. Even if you have never worked with a distribution plotted in this way, you are still familiar with this average, which is the everyday use of the word. If you doubt that, think of a distribution of four men, two whose heights are between 66.5 and 67.5 inches and two between 68.5 and 69.5 inches. We would then have just two points on our curve, and the average would be $(2 \times 67 + 2 \times 69) \div 4 = 68$ inches.

The second is the *standard deviation*, which is a measure of the width of the distribution and which everyone calls σ (sigma). It would take us into too much detail to define σ and show how it is calculated; for our purposes it will suffice to say that for many more-or-less symmetrical distributions, about 2/3 of the population lie within $\pm\sigma$ of the mean, and 95% lie within $\pm 2\sigma$ of the mean. For our purposes, it will be enough to remember that this σ measures the degree to which the data are spread out. If $\sigma = 0$, there is no distribution; all the data give the same value. If σ is very large compared to the average, the data are spread out so much that the average has little meaning.

Before going on, however, I should warn you that there is no law that says that distributions will have the familiar bell shape of Fig. 12-1. And it is dangerous to assume that a particular distribution has this shape if you are going to perform any mathematical operations using it. The deviations from this shape are especially obvious if the distribution exhibits a mean and a standard deviation that are comparable in size. Consider, for example, the population of some thousands of telephone calls. I plot in Fig. 12-2 the number of calls of a given length (say within ± 2 seconds of the length l) vs. the length l of the call (in seconds). Here $l_m = 160$ sec and $\sigma = 120$ sec. Even for this distribution, which is not at all bell-shaped, σ is still a useful measure of the width, the *dispersion*, of the distribution and is calculated in the same way, but the "2/3" and "95%" are no longer good approximations.

Another useful measure of a distribution is the *median*. This is the quantity such that half of the population is less than it and half more than it. In Fig. 12-1 it appears that the mean and the median are about equal, and this is true for many distributions found in nature. On the other hand (and especially when the mean and the standard deviation are comparable in size), the median can be quite different from the mean, as in Fig. 12-2. Much confusion about family incomes arises by confusing means and medians; the mean family income of a population of families is virtually certain to exceed the median substantially because of the effect of some very large incomes (although of very few families) in the "tail" of the distribution on the high-income side.

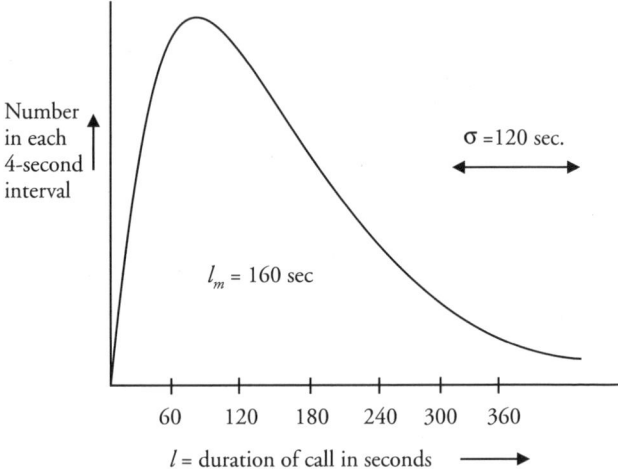

Figure 12-2. Distribution of the lengths of telephone calls.

Another descriptor that is occasionally seen is the *mode*. This is the value of the measured quantity that gives a peak in the curve. There is not much you can do with it since it does not appear in any of the mathematical operations applicable to distributions. The main reason for mentioning it is to warn you that the value of the variable that gives the maximum number in the population has very little significance. Another reason is to warn you that you may encounter "pathological" distributions that are *bimodal* (having *two* modes), and we shall encounter one soon.

But to return to a much more powerful concept, the language of σ as a measure of the dispersion (width) of a distribution pervades science and engineering. By extension into colloquial speech, you will hear a scientist describe a colleague's political views as "2σ to the right of Bill Buckley," meaning that he occupies a very lonely position on the far right. A prominent manufacturer has set a goal of "Six Sigma" quality, which means about three defects per million measurements.

Weather forecasters have an especially sensitive problem when they are forecasting imminent bad weather: How hot will it be during hot weather? How much snow will fall when a winter storm is on the way? After many years of listening to them (but not interrogating them), I am convinced that their usual practice is to state a forecast that is about σ to the bad weather side of what they really expect to happen. That is, suppose they expect a 4-inch snowfall with a probability of about 2/3 that it lies between 2 inches and 6 inches, thus with a σ of 2 inches. Then they publicly state a forecast of 6 inches, that is, σ to the bad weather side of the expected mean. If the snowfall then turns out to be only 4 inches (or even 2 inches), no one is likely to complain; and it is very unlikely that it will be 8 inches (2σ above the real prediction of 4 inches), at which point the screaming would begin. This may not be what they do, but I have noticed that weather is usually better than predicted, and I think it is likely that if I were a forecaster that is what I would do; my excuse for what would really be timidity would be that people should be warned of especially bad weather, even if it is unlikely.

Good advice about speaking to security analysts and the press about the performance of your company is to say as little as possible. If you believe you have to say something, then the classical advice is to "underpromise and overdeliver." But how much to underpromise? One σ is about right. Then, as in the weather forecasting, if the news is better, complaints will be minimal, and it is unlikely that the news will be worse.

One population can be described by many different distributions; the data can be distributed along different axes, as in the work of Chapter 6. For example, the population of men we discussed earlier can also be distributed according to their weights (instead of heights) as is given in Fig. 12-3.

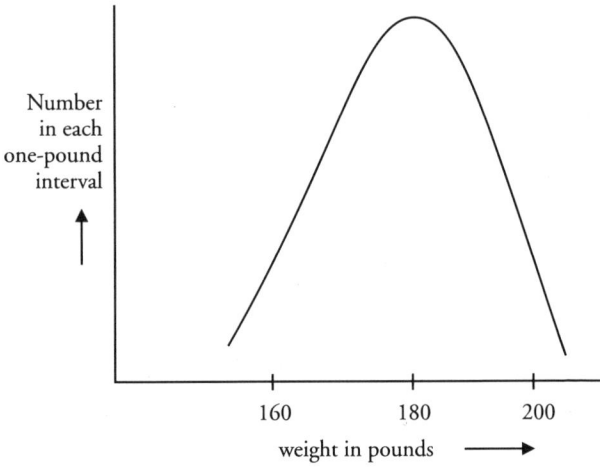

Figure 12-3. Distribution of the weights of a population of men.

Or we could distribute this same population according to the number of years of their formal education, Fig. 12-4a. For some populations, this latter distribution might be bimodal, as in Fig. 12-4b, in which the prevalence of high school graduates with twelve years and college bachelor's degree holders with sixteen years produces a distribution with two local maxima, two modes.

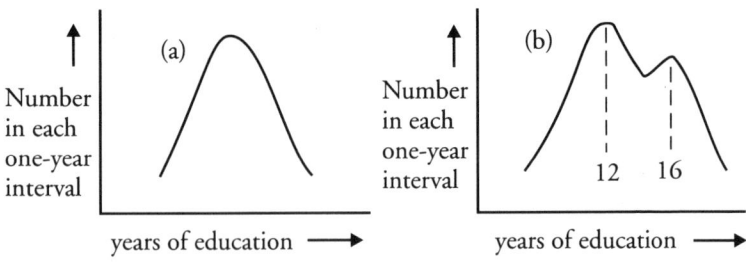

Figure 12-4. Distributions of the years of education of two populations.

If the same population is distributed according to different variables, the exceptional point in one distribution may or may not be exceptional in another. Sometimes there is a relation: A very tall person (say 2σ above the mean) is likely to be heavier than the mean weight, but not necessarily exceptionally heavy. Sometimes there is no relation at all: The person who is 4σ above the mean in mathematical ability may be just average in tennis performance or in knowledge of politics. It is often difficult for the person who knows she is exceptionally good in one respect to appreciate that he or she is not especially qualified in other ways; realization that the halo is not universal can be an uncomfortable process.

Sigma is a measure of dispersion, and you must be very careful if you attach any additional significance to it. Some investment advisers plot total return (price change plus dividends) from individual common stocks vs. the dispersion σ of the price of each stock over some period like the preceding year. For the investor who is averse to fluctuations or easily frightened, this can be a useful way of portraying the trade-off between yield and stability. But sometimes this σ is referred to as "risk," and the plots of yield vs. σ are claimed to be yield vs. risk.

For the pension fund, endowment, or even patient individual investor, this is misleading; the client can average over his portfolio and over time. Risk really is the danger of deteriorating management, burgeoning competition, or continuing to manufacture buggy whips when the horseless carriage is taking over. If this σ measures risk at all, it is the risk to the investment adviser, not to the investor; an impatient investor may discharge the adviser after a couple of quarters with down fluctuations. The conservatism engendered by avoidance of the risk of rustication may be one of the reasons why professional investment firms do not perform, on the average, better than the dart board or index fund.

Up to now we have considered statistics as descriptive of populations. As long as we are careful about specifying exactly what the population is, make good measurements, and include *all* the data, we cannot get into serious trouble with descriptions.

But the power, and the danger, of statistics comes when we *sample* populations and go beyond descriptions to use statistics for the *control* of processes and for drawing *inferences* about the populations sampled.

Suppose we take a sample of 10 men from the population of Fig. 12-1. We may get Fig. 12-5, with a mean H of 66.9 inches and a standard deviation s of 3.0 inches. Or we may take a second sample from the same population and get Fig. 12-6, with H = 68.2 inches and s = 2.2 inches. (We use new symbols H and s for the sample, since they will generally differ from the h and σ of the population; I have not explained how to calculate s, but you can verify just by "eyeballing" the samples that the s of the second sample is somewhat smaller than that of the first.) Or we may get many other possible samples. For some purposes, such small samples might suffice, but if we were deciding how many of each size of men's suits to stock in a store, it would probably pay to get more data, even though collecting more data clearly incurs costs.

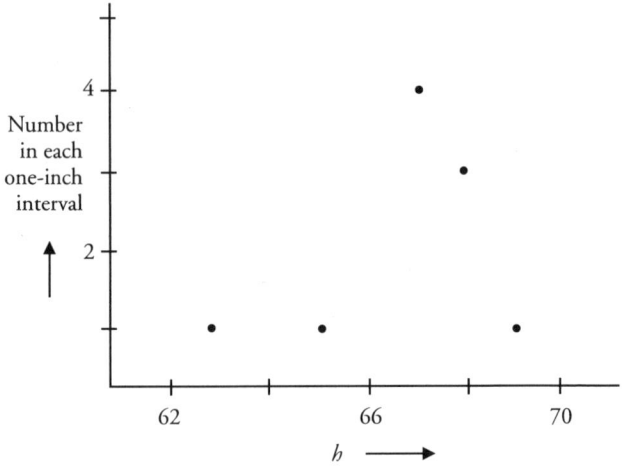

Figure 12-5. One sample from the population of Figure 12-1.

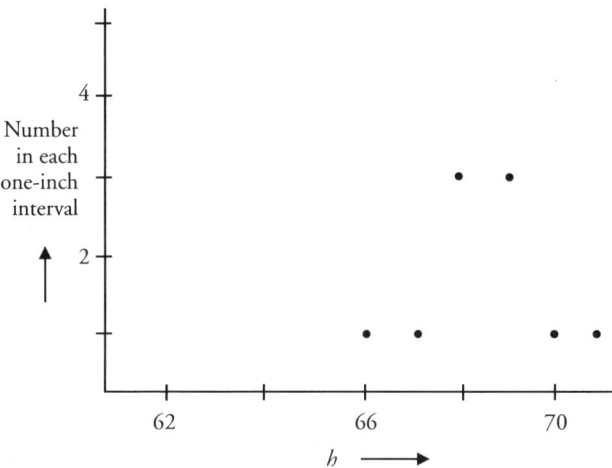

Figure 12-6. A different sample from the population of Figure 12-1.

If we had not known that these two samples were from the same population, we might have been tempted to assign some significance to the differences between them, for example, the difference between the two means. But from the information I gave you about how the figures were derived, we know that although the means H_m of the samples are different and each is an approximation to the mean h_m of the population, there is no significance in the difference between the means of the samples.

How do we take a proper sample? It is not easy, and sampling is fraught with errors. If you take the men who walk past a certain point in an interval of time, tall men may walk faster or be more likely to be out walking than short men. If you take all those whose surnames begin with "D," there may be an ethnic bias in which more "D" surnames than average include short people from a particular ethnic background. If you use the last digits of the individual's telephone number and a table of random numbers, you exclude all those without telephones, which may bias the sample.

Sampling is an art as well as a science, as is evident from the enormous difficulty experienced in obtaining a representative sample of how people intend to vote in a presidential election (and in that task, of course, there is the additional uncertainty that they in fact will vote and will vote the way they said they would vote).

It is extremely easy to err in sampling, frequently without thinking about it. You often hear: "The airlines claim that their load factors (percentage of seats filled) are small, but nearly every flight I have been on recently was almost full." Well, your chances of being on a well-patronized flight are higher than on a nearly empty one, roughly in proportion to the number of seats filled. At the extreme, if the airplane were empty, there would be no chance that you would be on it, and thus it would never appear in your statistics. (This technique of looking at the extreme case is a common help, which we will examine more closely in Chapter 16.)

Mistakes in sampling can be intentional as well as accidental. A television news show reported on the maintenance experience of a new helicopter model. They reported that all of maintenance workers they interviewed reported problems with its reliability. Well, how did they choose the sample? Did they carefully select a random sample? If they did, almost any statement beginning with "all" would be unlikely. Or did they post a notice: "Any maintenance workers who have had problems with reliability are invited to be interviewed"? They do not say, and it is well that they do not claim to be a documentary.

The questions of sample size and quality become acute when you are attempting to use a sample to infer quantitatively a property of the population. For example, suppose someone had taken the sample Fig. 12-5 from a population of 1000 people who were either all men or all women. Suppose he then had taken the sample Fig. 12-7 from another population of 1000 people, again either all men or all women. Suppose we have been given the samples without knowing how they were taken. Are we justified in concluding that these are different populations, the second being significantly shorter than the first?

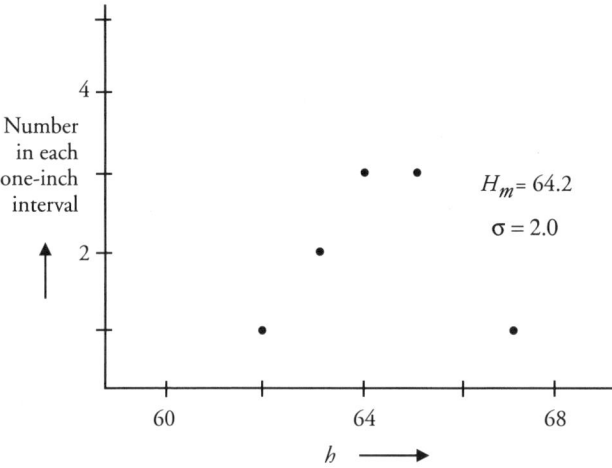

Figure 12-7. A sample from a different population.

That is the type of question that the tools of mathematical statistics were designed to answer, and by sophisticated methods the statistician can draw correct conclusions without unnecessarily expanding the sample size. Teasing out reliable conclusions from statistical samples is a most worthwhile occupation, but it should be done only by experienced and strictly honest individuals.

Has a change in the supplier of a part caused a significant change in some measure of the quality of your product employing that part? Large samples to answer this question will be costly, and it may even be impossible to get a large sample, since, for example, a limited sample may have been taken of the original product all of which has now been shipped. Here is where you call in a good statistician.

I should emphasize again the crucial importance of care in sampling. In the example just now considered, perhaps interest in the question developed late and the sample of the product was all of the product shipped on the last day the "old" supplier's parts were included. It would be a treacherous procedure to compare this sample with the product produced on the first day the "new" supplier's parts were included. First, the two samples should be from the same day of the week, since in many industries (and best documented in automobile assembly plants) the day of the week affects the quality of labor. Second, the product with the "old" part was presumably well along on the learning curve, and the product with the "new" part is just starting on that curve as it applies to incorporating that part into the product. And there are many other traps. The danger is not in using mathematical statistics per se, but in attaching too much significance to conclusions or acquiring too much confidence in mathematical operations based on poor or inadequate sampling.

Additional traps enter when human behavior is part of the operation being tested statistically. The most famous example of this is the "Hawthorne Effect." The American Telephone and Telegraph Company and its then subsidiaries Bell Telephone Laboratories and Western Electric Company were pioneers in the applications of probability and statistics. Determining what the capacity should be of long distance telephone lines and switching equipment that are to be installed depends critically on the statistics of telephone calls and the estimation of the probability that busy signals will confront subscribers. With the powerful statisticians these companies had on board for this purpose, it was natural that the manufacture of telephone sets at the Hawthorne, Illinois, works of the Western Electric Company would be analyzed statistically and that experiments would be conducted using statistical quality control.

A population of workers was used as a group to experiment with different assembly methods. Every change, including the "change" back to the original method, produced better results! Statistically significantly! What was clearly happening here was that better human performance was being obtained when the workers realized that management was paying attention to them and taking their work seriously and when they identified themselves with an "experimental group." (Recently some criticism of the analysis of the Hawthorne experiments has appeared, but the basic conclusion has been well confirmed by other experience in the intervening years.)

The weak reason for mentioning the Hawthorne Effect is to caution you about the extra care required to gather meaningful statistics when they involve human performance. The strong reason is to suggest that you can take advantage of the Hawthorne Effect to improve performance, not only in manufacturing but also (and probably especially) in relations with customers and in the service industries. Your employees are likely to perform better if they believe they are part of an experimental group or if they believe you are paying attention to them. I am not proposing that you fake experiments. But it is not fraudulent to let people know that you are interested in the conditions under which they work and the instructions and incentives given them. And anyway there is often much to be gained by varying working conditions.

Since we have moved into the realm of the behavior of workers, I digress at this point to introduce another piece of scientific language that is useful here: *boundary conditions*. If, for example, I am solving equations to determine the strength and direction of the electric field within a metallic enclosure (or the reflection of a light wave at a metal surface), it is necessary to impose the boundary conditions that the field is zero at the electrically conducting surface (it is not actually zero, but it decays so rapidly to zero inside the surface that for most purposes no error is made). In chemistry I start the analysis of a reaction by specifying quantities of the reactants and specifying temperatures and pressures; after a long enough time the concentrations of products and the new temperature and pressure will be the same throughout the container. These boundary conditions are an important part of any scientific calculation, as important as the equations. It is not much of a generalization to use this language when considering human performance and the behavior of workers.

You as an executive are already familiar with boundary conditions but perhaps not with the name. Anyone with responsibility to a group of employees or customers operates within narrow constraints. Of these the legal constraints are the most obvious, but if you are to avoid giving hostages to the media you must impose an even more restrictive self-restraint. The Ralph Naders of the world, with no responsibilities, can use flamboyant language to get headlines on page 1 and financial contributions and can move opportunistically from issue to issue. As you know, your correction in measured language appears three days later on page 15.

To revisit behavioral experiments for a moment, I have a theorem about experiments in education, such as a different high-school mathematics course or a new college curriculum. It is that "all educational experiments succeed." The theorem may not be exactly and literally true, and I cannot give you a proof. But my (misused, unscientific) statistics support it. The reasons are probably the following: (1) Rarely does an educational experimental program have a *control* group, identical in every way except in the experimental point being tested; and (2) Almost all educational experiments are evaluated by the people who promoted them. If you conduct experiments in your company, it would be well to note this experience and to insist that evaluation be by disinterested groups.

But I must return to statistics. Statistical process control is becoming more important every year and deserves even more rapid and extensive employment. It played a key role in the Japanese quality miracle of the 1950s and 1960s. A new wave of utilization is now under way under the heading of Taguchi methods, which include sophisticated experimental protocols testing several variables (e.g., in a manufacturing process) in one experimental cycle.

The older process control is illustrated in Fig. 12-8 in which the number of parts is plotted against some key variable. The variable might be a dimension, for example, and parts lying outside the limits will not fit or will show ugly gaps (as in automobile door fits). Or it might be the voltage of an electricity supply, and too low or too high voltage will stop operation or damage the works. Sampling theory was highly developed so that the sampling was as unintrusive and inexpensive as possible yet warned when the machines were turning out unacceptable products.

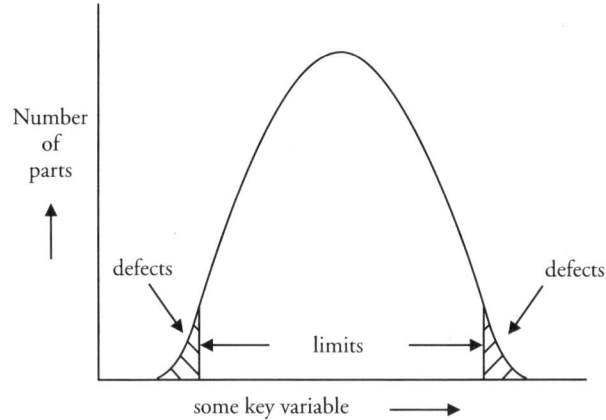

Figure 12-8. Traditional process control.

Comparison with the newer approach can be illustrated by the "fit" (the gap at the edge) of the trunk (boot) lid on an automobile. In Fig. 12-9 the old practice was to specify limits within which the gap must fall, perhaps 0 inches on the low side (the lid would scrape or not even close if it were < 0 inches) and $3/8$ inch on the high side. Statistical control would then be imposed on all the processes affecting this gap: the dimensions and assembly of the pressed steel parts surrounding the lid; the dimensions of the die for pressing the lid; the metallurgy of the sheet steel to be formed into a lid (which affects the "spring back" after removing the lid from the die); and others. Only a few parts on the little tails outside these limits would be rejected, but the resulting cars did not appeal to demanding customers; many cars exhibited gaps right up to the "acceptable" limits, and customers were not impressed.

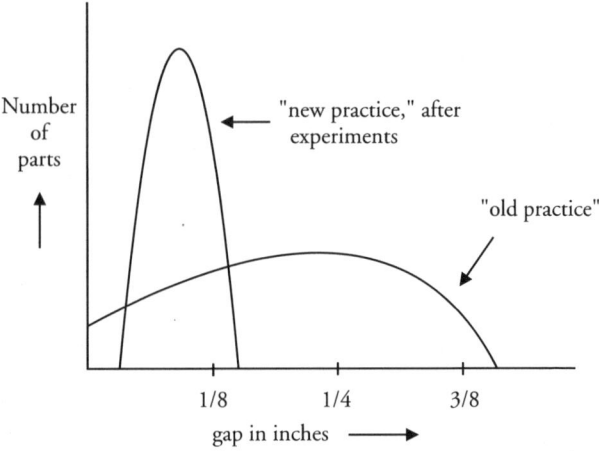

Figure 12-9. Modern process control.

The newer approach is to ask first: What do I really want the gap to be? Not zero anywhere, since a line would show somewhere and if the line were interrupted by a tightly fitting region it would make an unattractive fit; also a zero gap somewhere would render obvious any failure of the meeting parts to be flush. A good target is 1/16 inch. Then statistically controlled experiments are undertaken to *narrow* the distribution and even to change its shape, so that *no* fits are < 1/32 inch and as few as possible are > 1/16 inch. As preproduction continues along the learning curve, these targets are achieved. As wear occurs in the dies or assembly jigs, the changes that cannot be prevented without undue expense are compensated by changes intentionally introduced elsewhere in the process.

Similarly, equipment such as radars that are subjected to a variety of environments and have parts that wear in use are produced under statistical control that builds "robustness" into the product. The distribution of characteristics of the radar (e.g., the minimum voltage required of the electricity supply) is narrowed and positioned such that in use over many years the radar will remain operable.

Another approach to employing statistics was, in fact, introduced at the end of Chapter 5 where we were considering curve fitting. The simplest form, you will remember, was fitting a straight line to a group of data points. The most common way of fitting a line, least squares, is to draw the line such that the sum of the squares of the deviations of the points from the line is a minimum.

When least squares line-fitting is applied to statistical data it is a special case of *regression analysis,* a statistical treatment of data to tease out relations between variables. This least-squares approach is "linear regression" and with a particular criterion for bringing out the relation between the variables. "Multiple regression" involves several variables and an attempt to learn the effect of each; we touched on it (without naming it) in referring to Taguchi methods.

Regression methods are powerful and helpful in process control. But they are also commonly used to claim to establish the *cause* of some effect. In this role they are extremely dangerous unless used carefully by statisticians who are scrupulously honest and who have a good grasp of the processes under study (they are not just looking at statistics from a remote position, never observing and critically reviewing the data collection).

Let me give a trivial illustration that would naively lead to such an outlandish result that you would not be misled. In Fig. 12-10, I plot the power of an old steam locomotive as a function of the height of a common housefly above the floor of the cab. I can draw a curve (not quite a straight line, but that is not important) through the points, but who would believe that the fly *caused* the power? You see, I "forgot" to tell you that the fly was sitting on the throttle lever!

This (absurd) example illustrates the profound difference between statistical *correlation*, one variable moving more or less in registry with another, and *causation.* Even though two properties or variables are closely correlated, even if one is almost exactly proportional to the other, that does *not* constitute evidence that one causes or is caused by the other.

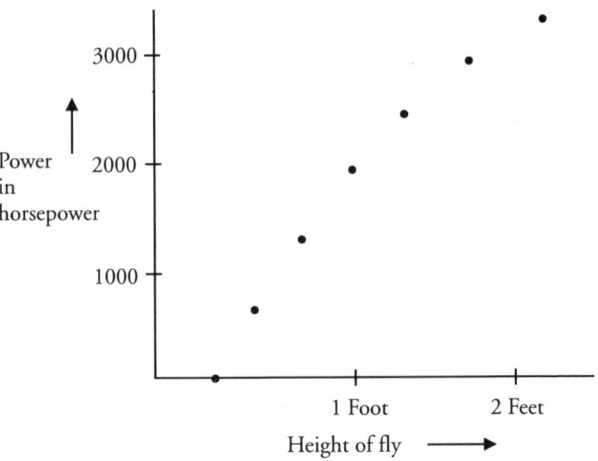

Figure 12-10. Locomotive's power as a function of the height of a fly.

In an article in *Nature*, Helmut Sies presented data that reveal a very high correlation between the number of babies born in Germany from 1965 to 1980 and the number of pairs of breeding storks during that same period. Do storks bring babies? And in the most recent military draft, Walter Oi found that the incidence of exemptions for physical reasons, mainly asthma, was positively and strongly correlated with the individual's education and his parents' income. But surely we cannot conclude that money causes asthma or that asthma causes wealth!

This warning about correlation and causation needs to be heeded especially when data are plotted as a function of time. In the fly-locomotive example, the plot of the power as a function of time as the locomotive starts, accelerates, cruises, coasts, and stops would look *exactly* like the plot of the height of the fly as a function of time on the same time scale. You have not been misled here because of the absurdity, but if the quantities plotted vs. time were less transparent you could easily be misled. For example, complicated data like the rate of spawning of fish and the oxygen concentration in water can have plots vs. time that "show a strong correlation" with plots of the introduction of particular chemical or physical pollutants at particular places.

But the giant step to identify any one of these as the *cause* of the decrease (or increase) in spawning rate needs far more evidence than statistical correlation. Additional variables (tides, currents, time of day, weather, climate, and others) are almost certainly at work, and both the introduction of pollutants (e.g., heated water from a nuclear reactor) and the spawning of fish may be responding to them in the same pattern in time.

You may have wondered why I called out "honesty" a few pages back. Let me show why strict honesty is demanded by considering the common situation wherein the statistics tell the truth but not the whole truth. Examples occur every day, but one I remember particularly was of a "scientist" (that is what the newspapers called him) who "showed how to predict earthquakes." He published a paper showing a strong correlation between the occurrence of earthquakes in one county of the Central Valley of California and the rate of egg production in a Central Valley city 24 to 36 hours before the earthquake. The correlation was so good that, by perfectly appropriate and routine statistical methods, he calculated that the probability that such a correlation could occur by chance was 1/1000. That is, in only one case out of a thousand like it would the result occur if chance alone was at work.

But what the "scientist" did *not* tell us was that he had examined over a thousand possible correlations, involving more than twenty cities and rural regions, ten different time delays between the "anticipatory" data and the earthquake, and many kinds of data (egg production, milk production, temperature, barometric pressure, and the like). It was therefore to be *expected* that totally by chance one (or even more) of these combinations would yield the impressive correlation reported. This correlation was therefore to be expected solely by chance, and there was no predictive power whatsoever in his "discovery." He was not honest with us and did not tell the *whole* story, and without that, the part he did tell was profoundly misleading. We must have not only the truth, but the whole truth.

Particularly close attention has recently been attracted to the compensation of the chief executive officers of large corporations. Salary administration in a corporation or other organization is aided by systematic data collection, statistical analysis, and comparison with the analysis in similar organizations. Salary administration consultants and firms provide a useful service by enabling corporations to participate in comparison groups without releasing confidential data.

This approach is fair and useful for 95% or 98% of the employees, but it has serious defects when statistical measures are applied to the top few percent because the sample sizes are small and the comparability of positions in different corporations is far from assured. Salary administration consultants like to quote means, medians, standard deviations, and even more recondite measures to compare executive salaries, but although these give the impression of a science, they can be seriously misleading. I cite only two of the possible traps.

It frequently happens that the corporation has only a few (say two dozen) comparable firms, and none of them competes for exactly the same skills and experience. To illustrate this situation, I have shown a *scatter diagram*, Fig. 12-11, in which the total compensation of the chief executive officer of each of 24 companies is plotted vs. the annual sales of the companies. The consultants have been helpful to provide such data, and although each participant recognizes his own point, he is not supposed to identify any of the others. A regression line could easily be calculated from these data, but what use could be made of it? I have not drawn it because as soon as it appears on the plot it acquires a life of its own and diverts attention from the underlying data. Even worse, sometimes the slope of this line will be provided by the consultants without letting us see the scatter diagram, on which two or three unusual points have a determining influence on the line. Responding to the derived statistical measures without seeing the raw data can be very treacherous. This is the first potential trap.

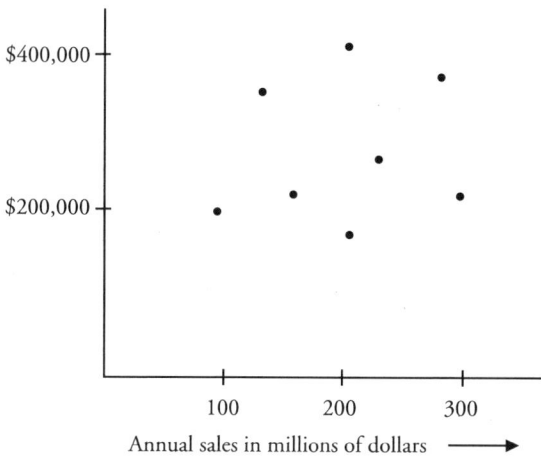

Figure 12-11. Chief executive officers' compensations vs. their corporations' sales.

What we need to do is to place our CEO on the chart, to use all the information we can collect about other CEOs near him on the chart (experience in the job, challenges peculiar to each corporation, track record of increasing shareholder value, etc.) and then to determine whether our executive is appropriately compensated. (Although the producers of data sets like these usually promise that the identity of each corporation will be kept confidential, it is always possible to infer the identity of some and frequently of most—in this example, we can look up the published figures for sales of these corporations and match them with the horizontal positions of points).

Scatter diagrams are especially useful as the first step in analyzing data. Many combinations of price and volume data, for example, can be plotted in scatter diagrams to explore the relations among them.

The second potential trap in our example is that salary consultants usually ask the chief executive officer or the board of directors of a corporation: Where do you wish to position your corporation's executive salaries with respect to your competitors? The answer often is: We wish to have better-than-average performance of our company, and therefore we expect to pay for better-than-average management. (I have never heard a CEO say that he expected average or worse!) The consultant then suggests that executive salaries be set at the "75th percentile" point of similar corporations of comparable size. That is, 75% of the salaries for comparable positions (e.g., chief financial officer) would be smaller and 25% would be larger than ours. He recommends that appropriate adjustments be made "to bring our salaries up to our policy."

Can you see what would happen if every corporation took his advice? The situation *explodes*, and even if the comparisons are made only once a year, each yearly cycle pushes up salaries handsomely. That this does not happen (although it certainly has had an effect) is a tribute to the skepticism and good sense of CEOs and boards of directors, helped by the circumstance that many corporations do not use salary consultant firms.

The social sciences (economics, political science, sociology, anthropology) necessarily rely heavily on statistics and statistical analysis and inference. The misuse of statistics, primarily by people outside these professions, has unfortunately tarnished the reputation of the social sciences. I can illustrate the intimate connections between statistics and the social sciences by introducing you to the late Bernard Berelson's Three Laws of the Social Sciences. Berelson, a distinguished demographer and a social scientist above reproach, jocularly suggested that all the regularities claimed by social scientists were just special cases of these three "laws." He suggested that as you read them you think of any law of social science to which you have been exposed—for example, "rich people have more children than poor people"—and decide for yourself the applicability of his description.

Berelson's Laws	*Connection to Statistics*
1. "Some do and some don't."	Collection of statistics is necessary since no law has language like "everyone" or "no one."

2. "But the differences aren't great."

Statistical analysis is required to discover meaning in data where the effects studied are small. A small signal is extracted in the presence of a great deal of noise.

3. "And anyway it is more complicated than that."

In pursuing one relation or effect, several others may interfere; or the differences among populations may obscure the effect studied.

In concluding this chapter, I list the major additions to your tool bag and some cautions in their use:

- Statistical methods are powerful and often necessary.
- The language of mean, median, and standard deviation can be useful in many situations even if you do not calculate them.
- Use of small samples and good mathematics saves money and is often necessary at any cost.
- Thinking in terms of distributions and of a population distributed according to various characteristics is a helpful tool.
- Process control by statistical methods is essential to quality products and services.
- Statistical inference is reliable only if a host of traps and fallacies is systematically avoided.
- You should be skeptical of the person (usually not a statistician) who is using statistics to "prove" something.
- You should use the raw data and scatter diagrams to monitor the honesty of those using statistical measures and to make sure these measures have real substance behind them.
- You should insist on learning the *whole* truth.

13

Trimming Things Up

Science proceeds by somehow discovering the important variables and the enduring relations among them, usually in the presence of a multitude of complicating effects. Engineering design likewise concentrates first on the essential functions and parameters. Quite frequently complicating effects are handled by adding small terms to the main calculation, and these small increments can frequently be calculated only approximately. Quick and easy approximation techniques are useful generally, and they can help us keep our attention on the main, important effect. Further, they provide a way of sorting the elements of a business problem according to the relative importance of the individual parts.

Consider, for example, the simple pendulum, a heavy metal ball suspended by a wire from a fixed point. Only after centuries of lack of understanding, and after the painful transition to the idea of *experimental* science, was this problem stripped of its inessentials and reduced to the two key quantities, the length of the wire and the acceleration of gravity. The period of the pendulum can be calculated from these two quantities alone, and the calculation agrees with experiment to a fantastic accuracy.

Of course, once this basic understanding has been achieved you can estimate "corrections" to the formula for the period. For example, the mass of the bob and the size and material of the wire determine the amount by which the stretch of the wire at the bottom of the swing exceeds the stretch at the ends of the swing. A correction can be made for this stretch; it is very small indeed for common and sensible pendulums. Temperature, barometric pressure, and even the color of the bob have tiny effects. If one had been preoccupied with a truly exact theory including all these corrections, one would at the very least have greatly delayed one's understanding of the basic phenomenon.

This situation is well known to you as an executive: The ranking of the relative importance of issues, problems, and opportunities is the primary responsibility of the executive, than which nothing else is more important. Once ranked, you can concentrate on the principal issues and add the corrections later. For the executive, this is an art, not a science, but an art buttressed by quantitative considerations.

It is also mostly an art for the scientist, too, although there are some mathematics and mathematical language that are helpful. I turn now to introducing these, which also introduces an important way of simplifying problems through the use of approximations.

It will be easier if we start with a specific example, Fig. 13-1. Suppose I am going from A to B, a distance of 10 miles, but instead of proceeding in a straight line I go through a waypoint that requires me to deviate 20° from the straight path until I have covered half the distance and then, of course, turn back toward B. How much farther do I go than if I had proceeded directly?

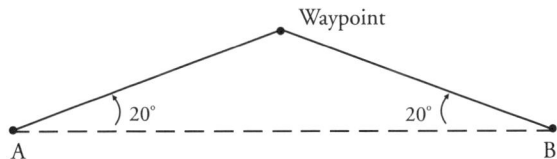

Figure 13-1. Deviating from the straight-line path.

You will doubtless remember that the definition of the cosine of an angle *x* is obtained from a right triangle as shown in Fig. 13-2.

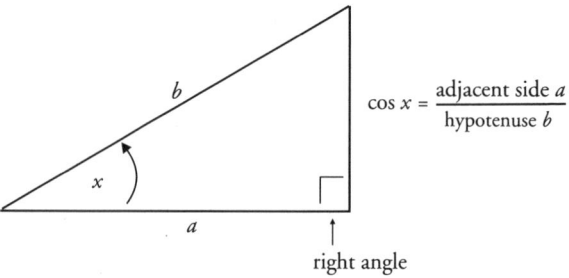

Figure 13-2. The cosine of *x* is *a*/*b*.

In our problem, *a* = 5 miles and *b* = 5/cos 20°, and the additional distance we require is 2*b* − 10 or 10(1/cos 20° − 1). Now we could look up cos 20° in tables or get it directly from many handheld calculators, and get 0.9397; and then 1/cos 20° = 1.0642. The additional distance is then 10(1.0642 − 1) = 0.642 miles, a 6.42% increase. But we really do not care about the answer and we do wish to learn, and so we will set about seeing how to get the cosine of 20° by an approximation technique.

First, to work on that 20°: Measuring an angle in degrees is an arbitrary, unnatural process; a full circle *could* have been called 100° or 24° or anything else as well as being called the conventional 360°. The *natural* unit of measurement of an angle is the *radian*, as illustrated in Fig. 13-3. If you think of an angle as being swept out by a radius turning through that angle, the radian measure is (arc length) ÷ (radius). The correspondence with degrees is apparent: 2π*r* is the arc length for an entire circle, and so 2π radians equal 360°, or one radian equals 57.3°. Our 20° is thus (20/360) × 2π = 0.349 radian.

Actually, of course, we are more likely to start a problem like this with radians rather than degrees; if we simply eyeball the map showing A, B, and the waypoint (Fig. 13-1), we could estimate that the arc length of the angle in question is about a third of the radius, and so the angle is about 1/3 radian.

Science and engineering thrive on natural measures. We shall encounter natural logarithms later in this chapter. Quite fre-

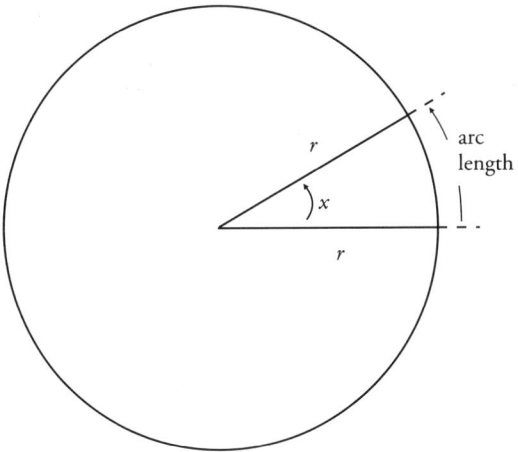

Figure 13-3. The radian measure of *x* is the arc length divided by the radius *r*.

quently identifying the natural or characteristic units helps us to think about or even to solve a problem. If you have ever had a house or cottage that relied on an inadequate well for its water supply, you will appreciate that the natural unit of water consumption is not gallons but flushes; the well's capacity in flushes per day determines whether you have enough water and whether you can afford to have house guests. The natural unit of distance for business travel is not miles but is hours on a commercial jet; all jet aircraft speeds are about the same, and so one thinks of the separation of Chicago from San Francisco as four hours, not as so many miles. The time for substantial sociological changes, such as in race relations or equal employment opportunity, is properly measured in generations (not in years), and thinking in terms of years may actually impede the process by creating short-term processes that have to be reversed or that at the very least subtract resources that could have been applied to longer-range solutions.

To return to our problem, I introduce the concept of a *power series*, applied here to calculate the cosine. Here 2! = two factorial = 2×1; 4! = $4 \times 3 \times 2 \times 1$; and so on.

$$\cos x = 1 - x^2/2! + x^4/4! - x^6/6! + \dots$$

The label "power series" comes about because it is the sum of terms with increasing powers of x. Where did this equation come from? It is easily derived through the use of calculus, but I shall not do that here. (Later, in the next chapter, I shall introduce the language of calculus but not make any calculations.)

Our cosine of 0.349 radians is then

$$1 - (0.349)^2/2 + (0.349)^4/24 - (0.349)^6/720 + \dots$$
$$1 - 0.06090 + 0.00062 - 0.00002 + \dots$$

Note how rapidly this series *converges*. Note also that what we need for our problem is only the second term, 0.0609, which is very nearly the amount by which the cosine differs from 1.

So, after all this, you have learned that you go only 6% farther if you deviate 20° from the straight path. But the technique you have learned is, of course, much more useful than the answer to this (arbitrary) problem. By including the effects of additional terms in the series we can make successively more accurate approximations of what we seek. But series calculations are primarily useful when only the first couple of terms are needed to get a sufficiently accurate answer.

Before going on with series, I take advantage of the example we have been working on to point out a general property of paths. Let us use the same 20° but with a different problem. Suppose we are flying directly from A to B (Fig, 13-4) but there is a crosswind blowing at right angles to the direction in which we wish to go, and of a magnitude such that we must point 20° away from the direct line in order that our actual track made good is directly toward B. Note that even though the wind is orthogonal to our path (neither a headwind nor a tailwind), we must go 6% farther to reach our objective.

This example illustrates why, on the average, environmental effects like winds or currents are *adverse* when you are seeking

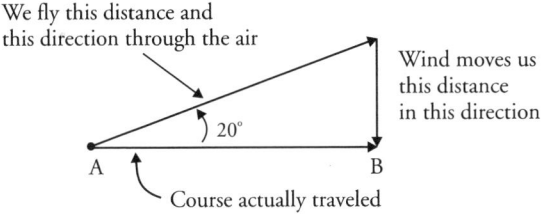

We fly this distance and this direction through the air

Wind moves us this distance in this direction

20°

A   B

Course actually traveled

Figure 13-4. Effect of a crosswind.

to attain a particular objective. If, on the other hand, you are simply flying or sailing around aimlessly, on the average, winds and currents are neutral, neither aiding nor impeding your motion. It is perhaps not too far-fetched to draw the analogy with administration: When you have a particular objective you are striving toward, the setting in which you operate is on the average adverse; if you are merely presiding over an ongoing operation with no particular aim, life is easier.

There is a related principle that governs negotiations: The person who cares the most loses. You must compromise and give up features you desire if you want the negotiation to succeed more urgently than does your counterpart.

Many functions in addition to the trigonometric function cos x can be expressed in power series. For example the sine of x, written sin x, is defined as the opposite side divided by the hypotenuse of a right triangle. The power series for sin x (Fig. 13-5) is

$$\sin x = x - x^3/3! + x^5/5! - \ldots$$

We encountered the natural logarithm in Chapter 3, but I did not define it there. It is just like the logarithm to the base 10 that we used there but it is to the "Napierian Base" $e = 2.71828\ldots$ Thus, the natural logarithm ln x is defined by the relation $e^{\ln x} = x$. The following useful

$$\ln (1 + x) = x - x^2/2 + x^3/3 - x^4/4 + \ldots$$

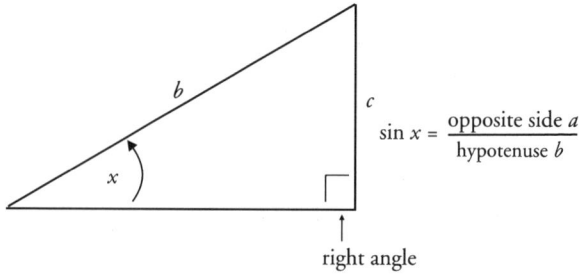

Figure 13-5. The sine of *x* is *c/b*.

series can be viewed as a way of defining this strange number *e*; had we used a different number, the series would not have the simple form with integers throughout.

Other, algebraic, functions can usefully be written in series form. For example,

$$1 / (1 - x) = 1 + x + x^2 + x^3 + \ldots$$

I actually made use of this series when in concluding the waypoint example, I in effect wrote $1 /(1 - 0.06) \cong 1 + 0.06$ (did you catch me doing it?).

The companion series is

$$1 / (1 + x) = 1 - x + x^2 - x^3 + \ldots$$

Another application of series may interest you, especially if you sail or fly: If your eye is *h* feet above the surface of the sea, how far away *d* is the horizon? Both *h* and *d* are tiny compared to the radius of the earth (3960 miles), and therefore our approximation technique works wonderfully, taking only the first terms. Fig. 13-6 defines the problem, and we turn the basic triangle on its side in the second figure (Fig. 13-7) to make it look more familiar. From the definition of the cosine of the angle *d/R*, in radians, of course, we get

$$\cos d/R = R \div (R + h) = 1 \div (1 + h/R)$$

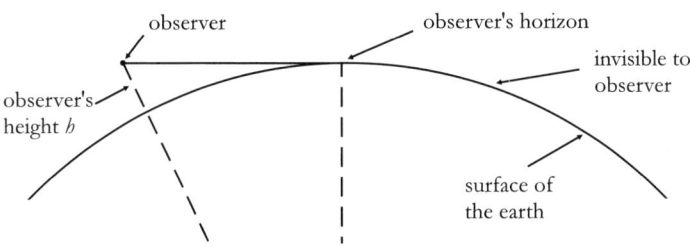

Figure 13-6. An observer a height *h* above the sea looking at his or her horizon.

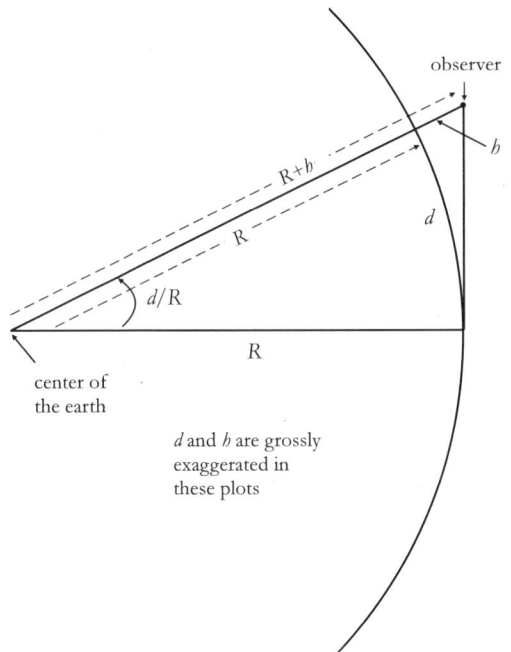

Figure 13-7. A replotted Figure 13-6.

But from our earlier series, $1 \div (1 + h/R) \cong 1 - h/R$
  And so,

$$\cos d/R \cong 1 - h/R$$

By using the series for the cosine we get also that

$$\cos d/R \cong 1 - (1/2)\,(d/R)^2$$

Therefore, equating the two expressions for the same cosine yields

$$1 - h/R = 1 - (1/2)(d/R)^2 \quad \text{or} \quad d^2 = 2Rh$$

We thus have the required result $d = (2Rh)^{1/2}$. To make this useful we express d in miles and h in feet and use the fact that there are 5280 feet per mile:

$$d\,(\text{miles}) = \{2\,(R = 3960\ \text{miles})\,(h/5280)\}^{1/2}$$

Since $2 \times 3960 \div 5280 = 1.500$ and $1.500^{1/2} = 1.22$, after all this we have a useful formula:

$$d = 1.22h^{1/2}$$

Thus if you are on an upper deck of a cruise ship 64 feet above the water, the horizon is $1.22 \times 8 = 10$ miles away. If you wish to see beyond the horizon to the light of a lighthouse 100 feet above the water, you can do so from a distance of $10 + 12$ (since 12 is $1.22 \times 100^{1/2}$ the light's distance to *its* horizon) or 22 miles. If you are in an aircraft at 40,000 feet, the horizon is $1.22 \times 40000^{1/2} = 244$ miles away. You may see "1.35" in place of "1.22" in boating books; refraction of light rays by water vapor in the air near the surface of the sea bends light a little in our favor.

To return to our general discussion of series, these are all *infinite* series, which means that you must in principle keep calculating terms on the right indefinitely to get the correct expression. And you must make sure that $0 < x < 1$, or the series "blows up" and gives nonsense. But as you can see from the way we have used series, we need not worry about these niceties since we shall always be working with an x that is very much less than 1; it is only then that we can calculate a term or two and have a good approximation to the function.

To show another example of the use of such series, let me return to the compound interest problem introduced at the end of Chapter 3. If an interest rate of x compounded for n years results in doubling the original capital, then:

$$(1 + x)^n = 2 \text{ and } \ln (1 + x)^n = \ln 2 = 0.693$$

(We have looked up $\ln 2$ in tables of logarithms or have obtained it from our little pocket calculator.) We next use the delightful property of logarithms, $\ln (1 + x)^n = n \ln (1 + x)$. And if we use the series for $\ln (1 + x)$ and take advantage of the smallness of x with respect to 1, then to a good approximation $\ln (1 + x) = x$. Thus $nx = \ln 2 = 0.693$, or if we express x in percent (by multiplying by 100), $nx = 69.3$. Thus we have derived the Rule of 69 (or 72), which we found useful in Chapter 3. (And, incidentally, by keeping the second term in the series we could show why "72" is a little better than "69" for interest rates in the range 5–10%.)

Convergence of these series is, of course, essential, and if we have a choice we choose the series representation that converges most rapidly.

I would ask you to think of the successive drafts of a document as the terms in a series. The first term, the "leading term," is the first draft. The second term represents corrections or additions to trim it up, and so on. In managing discussions with lawyers, accountants, and others, you must constantly try to make the series converge as rapidly as possible, so that fewer drafts are needed.

I once had an associate whose thought processes and mental discipline were so sloppy that when he was assigned the task of refining a second draft of a policy document, after working it over he converted it into a first draft! Clearly no convergence!

The leading term in a series often, as in the examples we have considered, tells most of the story. An individual can be viewed as a series, and such an approach can help us to work with him or her. For some people the leading term is caution, or courtesy, or brilliance, or generosity, or determination; for some it can be ambition or even greed. Identifying the leading term, and maybe even the relative size of some of the subsequent terms, can enhance the effectiveness of your interaction with the person. Series expansions are a good arena in which to consider what we mean by "negligible." A scientist never says that "[something] is negligible," but only that "it is negligible with respect to something else." We have in this chapter frequently neglected a later term with respect to an earlier term, and the series lets us easily estimate how good this approximation is. This process is typical of the scientific process of stripping a problem to its core and then examining what errors, if any, have been made by the simplification.

The term "order of magnitude" is rather loosely used in science. Frequently it means a factor of 10, as in "the annual budget of my company is three orders of magnitude larger than my personal budget." But the language can also be used to identify the terms in a series: "To first order, the cosine of x is simply 1; to second order it is $1 - x^2/2$." If you have carefully ordered the relative importances of the elements of a complicated question or problem, you may be able to say that one element is "only a fifth-order question," which means that you can safely forget about it unless you need high precision and everything else can be very accurately taken into account.

Boards of directors or trustees cannot substitute their management for the real management of a corporation or institution, except in crisis situations and then with great difficulty and risk. But what boards can and must do is "second-order management," in which they evaluate and monitor the quality of and choices of directions and priorities by the "real" management. They are accustomed to this role in functioning as audit committees, studying financial practices and performance at one step removed. As I suggested in Chapter 2, it would be well if boards would generalize this role and, in a second-order way, audit other normative features of corporate or institutional performance such as recruiting and retention of staff and quality and relevance of research and development.

"Second-order agreement" can be a very useful technique for resolving controversies and permitting management decisions in the face of basic disagreement among your associates. Suppose A disagrees with B and both have eloquently presented their arguments to you. You then ask A to present B's argument in a way that B will say, "Yes, that is exactly how I am presenting my case." And the same to B, presenting A's case. You have achieved second-order agreement. Frequently just the process of explicitly stating the other person's strongest arguments narrows or even removes the disagreement. And even if it does not, you are still in a better position to decide the question, basing your decision on agreed-to arguments with no obfuscation or poor communication.

When using approximation techniques to trim up and make more exact a particular position, it is important to choose the starting point carefully. Choosing the first approximation, the point from which trimming up will proceed, must be done with care. A procedure that I found difficult to counter was occasionally used in a friendly way by a highly respected associate. If the argument was not going his way, he would say: "Bob, as a first approximation, let's consider that you're wrong."

You have learned in this chapter the way the scientist proceeds from the basic understanding of the core of a question to trim up his understanding by successive approximations. You probably will not use the mathematical techniques of series expansions, but thinking about a problem by identifying the leading term and then the subsequent (minor) terms can be powerful and helpful.

14

Change Revisited

Some have suggested that I am a compulsive calculus teacher, and I guess they have a point. Calculus is at once a marvelous invention and the heart of useful mathematics, since it is the method and language of change and motion. Teaching calculus has been, in fact, one of the few unalloyed satisfactions I can claim. When a student suddenly catches the spirit and power of calculus,

> One sees the expression in some widened eyes,
> The *claritas* of sudden revelation,

in the words of the poet Anthony Hecht.

Although I shall resist making this book into a calculus text, I do wish to introduce the *language* of calculus and a glimpse of its applications, in particular the concept of the *derivative*. There will be no proofs or derivations! Rather than developing the mathematics of calculus, I shall rely on simple graphs to define the derivative and to explain some of its utility. Please do not be frightened until you have given it a try. As always, it will help if you use a pencil and paper to try things out.

I start with a straight line and define the *slope* in Fig. 14-1; here the slope is constant because the line is straight; we would calculate the same value of slope wherever we examined the line. The slope is also, by definition, the derivative of y with respect to x and is written dy/dx.

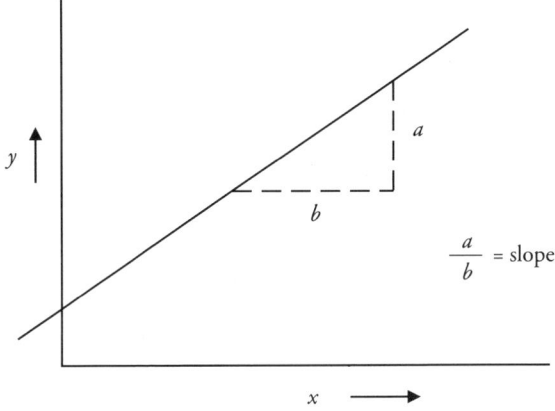

Figure 14-1. Definition of the slope.

Now suppose that instead of a straight line we have a more general curve (Fig. 14-2). We still define the local slope at each point as the slope of the line that just "kisses" the curve at that point, or, in more precise language, the line that is *tangent* to the curve at that point. This is the general definition of dy/dx, and the dy/dx of our curve is plotted at the right.

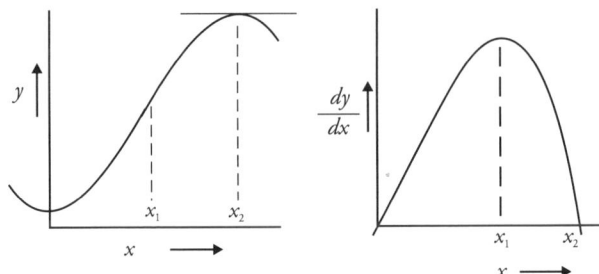

Figure 14-2. The local slope is the derivative dy/dx.

The derivative dy/dx is always the local rate of change of y with respect to x. As you see, it can be zero or negative. For this particular curve (which was made up arbitrarily), the derivative is zero at $x = 0$ and again at $x = x_2$. The derivative becomes negative (a slope downward to the right) for $x < 0$ and again for $x > x_2$.

We can also define the *second derivative* of y with respect to x, d^2y/dx^2, which is the rate of change of dy/dx with respect to x. (Similarly, third and higher derivatives can be defined, but they are not useful for our work.) You can already see why the *language* of calculus is convenient, since speaking of the "second derivative" is much less clumsy than speaking of the "rate of change of the rate of change."

You should not try to pry within these strange symbols like dy/dx or to assign significance to the "d's." You should just look upon the whole symbol as standing for the derivative of y with respect to x. I am sorry that these symbols seem mysterious, but I must use them because everyone uses them, and you may have learned some calculus using them. Historically, they came about because the dy's etc. originated in the deltas on curves like Fig. 5-9.

Let me make the derivative a little more familiar by the sketches presented in Figs. 14-3, 14-4, and 14-5. In each of these illustrations and at any value of x, you can sketch a local tangent, which is dy/dx, and verify the shapes of the curves plotted for dy/dx and d^2y/dx^2 (I have done this for dy/dx in Fig. 14-2 as an example). In Figs. 14-3 and 14-4 a is a constant. Although I have written in the actual expressions for these derivatives, I shall not use them or explain how I got them.

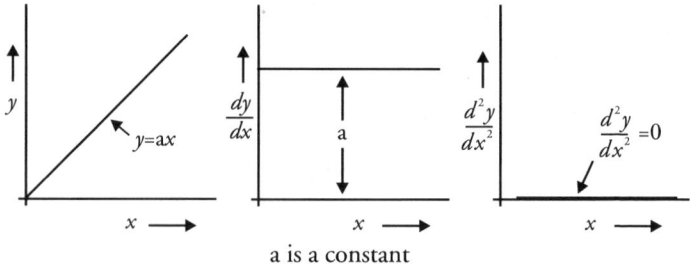

Figure 14-3. A straight line.

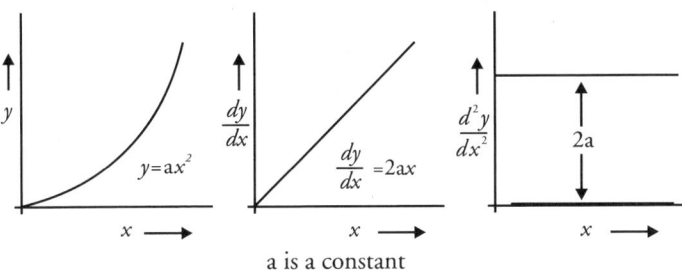

Figure 14-4. A parabola.

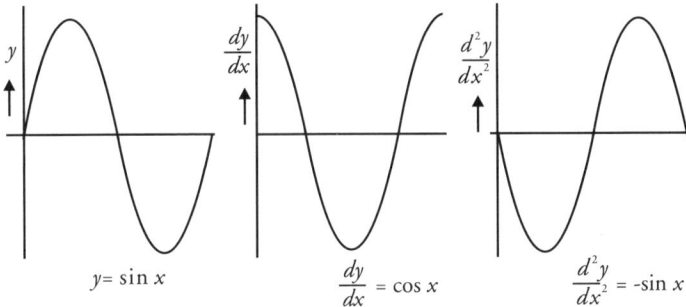

Figure 14-5. The sine of *x*.

The most obvious connection of all this to the real world is if *y* is a distance *s* and *x* is the time *t*. Then the first derivative = *ds/dt* is the *velocity* (the rate of change of distance with respect to time), and the second derivative *d²s/dt²* is the *acceleration* (the rate of change of velocity with respect to time). (It was, in fact, in order to deal with motion and to develop the laws of motion that Newton invented the calculus.)

To illustrate, let me portray (Fig. 14-6) the familiar experience of an automobile moving from one traffic signal to the next; both were red when we arrived (as is to be expected from Murphy's Law). I have placed these left to right in the customary order of the function, the first derivative, and the second derivative. My driving of the car, of course, is just the reverse: I specify the acceleration by the pressure of my foot on the accelerator and either manually shift gears or have them automatically shifted for me in response to the accelerator position and the car's velocity. Acceleration is thus the basic "input," and velocity and distance are the derived quantities.

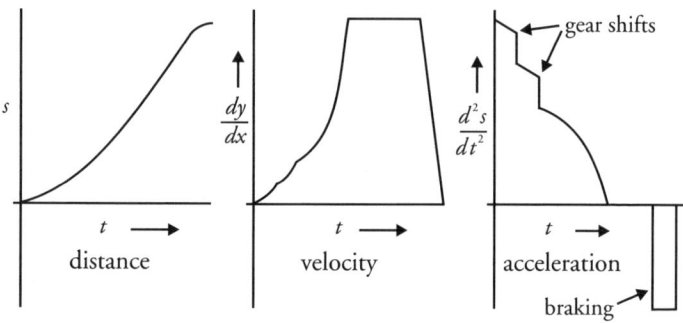

Figure 14-6. Standing start, cruise, and stop.

An important application of the derivative is in calculating maxima and minima. Suppose we have some function we wish to maximize, such as the gross margin on a particular manufactured product. If we can express this quantity as a function of one or more variables (such as the number made per shift or the payroll cost per shift), it may have a maximum (Fig. 14-7). If it does, the derivative will be zero at that point. In a simple, one-variable situation where we can plot it like this, we can eyeball the maximum and need not use any mathematics. The power of the calculus comes when many variables are involved and when we can calculate the derivatives as functions of these and set each equal to zero to get the optimum conditions.

Incidentally, you can easily observe that a maximum is characterized by a first derivative equal to zero and a *negative* second derivative (the slope is decreasing). A minimum also has a first derivative equal to zero but a *positive* second derivative.

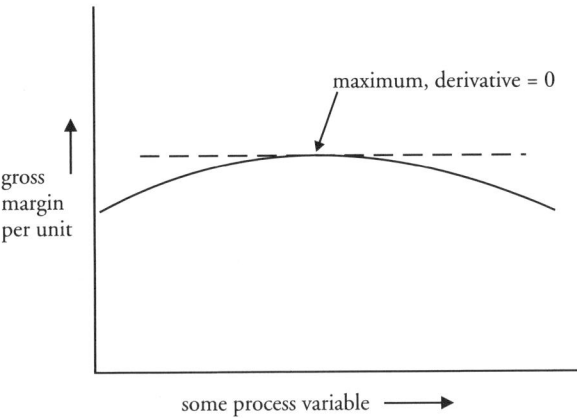

gross
margin
per unit

maximum, derivative = 0

some process variable ⟶

Figure 14-7. A maximum.

In optimizing, it is important to recognize that there may be more than one local maximum, as in Fig. 14-8. If one is not plotting (working only with mathematical functions) or if one unwisely restricts the range of the variables, he may "optimize" at the lower peak, which means a lost opportunity.

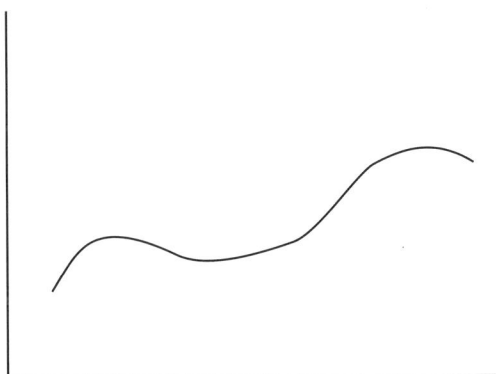

Figure 14-8. A function with two local maxima.

Again in optimizing, it is even more important to choose properly the function to maximize. Many of the problems of the American automobile industry stemmed from maximizing the attractiveness of cars in the dealers' showrooms rather than maximizing the utility to the buyer. A buyer's view of utility involves reliability strongly. Any deviations from perfect reliability impose enormous costs on the user; even if covered by warranty, the hassle of returning the car for repairs, doing without it for a time, perhaps returning it when the "repair" failed, and above all being stranded on the highway are costs that should be (and eventually were) included in the maximization process.

A more global illustration is the following: Suppose you operate a charitable organization and a direct-mail company proposes that they conduct a solicitation mailing for you. They, and you, estimate that the company's charges and costs will consume 50% of the contributions. You are upset at the inefficiency, but they point out that every dollar coming in this way is a dollar you would not otherwise get, so why worry about efficiency? If you optimize for your charity alone, they are correct. But those who give to you doubtless give to *other* good causes, and to the extent that they give to you (and cover those 50% costs!), they are unable to give as much to others. The over-all doing of good has suffered because that 50% went into nonproductive avenues. Thus the conclusion depends on which sum of contributions you maximize, whether you believe that your charity is so vitally important that you can let the total of all charity operations suffer.

This example prompts me to digress to explain a bit of language frequently used by scientists. If you are pedaling a bicycle or turning the crank on a winch, not all of the work you do appears as useful work in climbing a hill or lifting whatever is being winched. The rest goes into *friction*. In the bicycle case, there is friction in the pedal shaft, in the chain and gears, in the deformation of the tires, and in air friction as one speeds along. In the winch, there is friction in the gears and in deformation of the cable. We speak of the "frictional loss" in cases like this. It is natural, then, in the direct-mail example to say that 50% of the take goes into friction. Although not all overhead is friction, it requires eternal vigilance to reduce the friction to an irreducible minimum.

To return to faulty optimization, it is common and often disastrous in the case of public works projects for which the cost is shared by local, state, and federal governments. Frequently the real decision is left to local politics, and the argument is made there that the utility of the project more than justifies the "cost," but what is meant by cost is the cost to the local government only: "Using state and federal funds is just what all other communities are doing, and we should get our share." That may be, but the optimization process is not working, and disaster can ensue if there are heavy maintenance costs or if land is taken off the tax rolls for the project.

For a more direct use of derivatives, consider the head count in your company. If your business is growing, a positive first derivative is understandable. But you must watch carefully the second derivative. If you wish to increase the head count to a new value, as shown in the plot at the left in Fig. 14-9; your intention is to go from h_1 to h_2. You accordingly change the net hiring rate from zero as shown in the middle plot, that is, you make a positive second derivative in the head count.

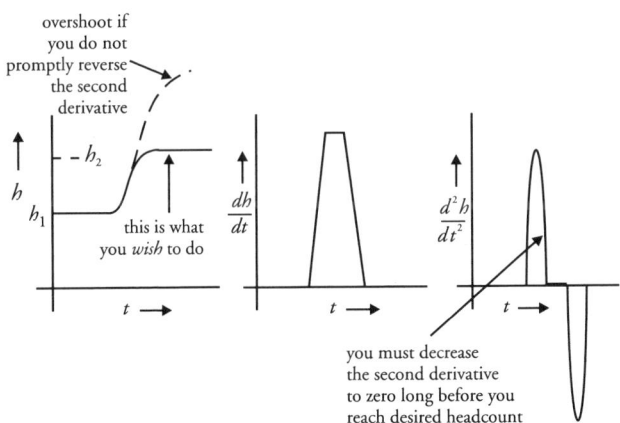

Figure 14-9. A head-count example of the dangerous second derivative.

If you are watching only the head count, however, you risk making a serious overshoot. In order to control the head count at the new level without overshoot, you must turn around the second derivative *early* in the growth process. Any continuing positive second derivative, however small, is a warning of trouble ahead, of an overhead growing out of proportion to needs. When you are looking at all three of these plots, your course of action is apparent, but if you were thinking only in terms of the head count, you might neglect the necessity of firm and early action to control the dangerous second derivative.

I have said from time to time that thinking in graphical terms is useful but needs not be reduced to paper. This problem of the dangerous second derivative is so crucial, however, that I strongly recommend plotting both the first and second derivatives as you proceed in your expansion process. It could save you your company.

An example of this danger comes from the history of medical education. A tremendous effort was made, with the injection of federal and state funds as well as investment and risk-taking by private schools, during the 1950s and 1960s to increase the rate of "production" of M.D.'s. Now demographic models can get pretty complicated, including as they must deaths, retirements, motion into and out of the country, and motion out of the profession. But the leading term in the rate of change of the number of practicing M.D.'s in the country is the production rate, the new M.D.'s per year; it constitutes most of the first derivative of the number practicing. As all this concentration was occurring on increasing the first derivative, the second derivative became appreciably positive. Because only the number practicing, which was still growing rather slowly and still showed (according to some criteria) a shortage, was tabulated and quoted in policy studies, attention was not applied to the dangerous second derivative. As a result, a serious overshoot in supply has occurred. (The effects of this are still unknown. According to conventional economics, the earnings of M.D.'s should decrease and medical costs should decrease. But medical economics is different, and more M.D.'s may very well find more things to do and increase medical costs.)

A quite similar situation prevailed with respect to the size of hospitals. For many years, expansion was heavily subsidized, and the subsidy continued because "there was a shortage of beds." Attention only to the function itself (the number of beds) led policy makers to continue a strongly positive first derivative and even a positive second derivative long after the second derivative should have become sharply negative.

The head count is the most obvious application to business of the reasoning associated with Fig. 14-9, but the capacity of production facilities is at least as important. You must turn around and reduce the rate at which you expand capacity long before you have reached the desired new capacity.

When the chief executive officer of a major corporation retires, his replacement has usually been identified for several years. No matter what characteristic of the company you plot (e.g., fraction of sales overseas, or average profit margin), the plot (Fig. 14-10) will show hardly a hiccough at the time of the changing of leadership. The function and even its first and second derivatives may be continuous. That is not to say that there may not be changes, but they will be made over a period of time and without sharp breaks. It is useful, if you are presiding over a change in CEO, to spell out explicitly which derivatives of which

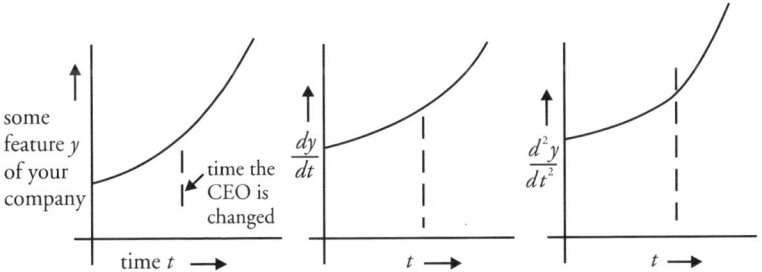

Figure 14-10. A change in CEO in a corporation.

aspects of the corporation you wish to change and how large a change to make.

For complicated reasons, the situation is quite different upon the retirement of the CEO of a college or university, Fig. 14-11. A national search is usually made, and the new CEO may have had no experience at the institution or sometimes even at institutions like it. The first derivative (and even more likely, the second) for many characteristics (e.g., relative attention to undergraduate and to graduate studies) can make a sharp, discontinuous break. This can result in damage, or at the very least inefficiency, but usually it is a benign injection of change. Change comes hard on a college campus; it has been said that changing a curriculum proceeds with all the speed of moving a cemetery. Thus even if some lost motion occurs, the change is frequently welcomed and is not likely to be too profound. A college or university is not brittle and easily shattered; it is a resilient structure primarily because of the continuity and leadership of the faculty but also because of the continued loyalty of trustees and alumni.

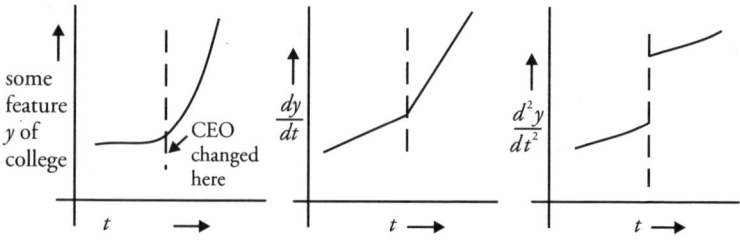

Figure 14-11. A change in CEO in a university or in a corporation in trouble.

The discontinuous behavior of Fig. 14-11 can be exhibited in a corporation if the corporation was in trouble and the new chief executive officer brought in from outside. Again, it is worthwhile to make these changes explicitly and to use the language of calculus to reason about them and to explain them to others.

You have not really learned any calculus in this chapter, but you should at least have learned the language for dealing with rates of change. Since you, the executive, are constantly immersed in change and challenged to manage it (rather than to let it manage you), ways of thinking about and portraying change are especially valuable. You will, of course be controlling the direction of the enterprise, but mastering the second derivative, the rate of change of the rate of change, is essential to controlling its size. The second derivative, like the accelerator on your automobile, is the key executive element.

15

Managing Operations

Operations analysis began in a small way with time-and-motion studies in manufacturing operations about 100 years ago. It grew steadily but slowly for 50 years and then received an enormous boost, both in the extent of applications and in the sophistication of analyses, during the military operations of World War II. Although it can be a complicated subject, with mathematical statistics in full bloom, there are also many contributions that operations analysis can make to your activities by using only simple applications of probability and statistics.

A famous example of operations analysis was the operations of British bombers against German submarines in the Bay of Biscay during World War II. Early in the operation, aircraft maintenance was evaluated (and medals and reprimands dispensed) on the basis of the number of aircraft in condition to fly. Clearly, keeping aircraft on the ground helped to increase this rating, this measure of performance. By changing to the measure of the number of missions flown, the effectiveness of the whole operation was substantially enhanced.

This example recalls the choice of what to optimize that we faced in the preceding chapter. Both are examples of the risk of what is usually called *suboptimization,* which is the process of optimizing (usually maximizing) some *part* of a complicated operation. Operations analysis demands that we study a *whole* operation, its objective, and the appropriate measures of its success. Although this is an obvious enough concept, it is extremely easy to slide into suboptimization, especially if different people or groups are responsible for the various parts, and these units have their own pride and morale based on what happens to their parts.

The elements of operations analyses are the identification of objectives, the specification of limitations (boundary conditions), the devising and execution of experiments, the application of mathematical statistics and economic analysis, and the comparison of predicted outcomes as conditions (parameters) are varied.

Operations analysis is now applied everywhere in military operations but has not penetrated nearly as far as it should into civilian life. To be sure, PERT charts and analysis are common in the construction industry and in major aerospace projects. But every operation that involves more than a single individual could profitably employ the attitude and statistical treatment of operations analysis. For example, the triumph of FedEx over the U.S. Postal System occurred in large part by the application of operations analysis, although imaginative management and the freedom from congressional and other government limitations must surely also have been important. Even such disparate activities as scheduling nurses in a hospital, establishing the prices of tolls in automobile tunnels, and creating emergency evacuation plans for highway accidents involving hazardous materials can profitably use operations analysis.

One small piece of operations analysis is the concept of the *rate-limiting step*, a concept imported from chemistry. The rate at which a chemical reaction proceeds is, of course, important in chemical processing and in the application of chemicals. The reaction rate may be zero, for example, in the breakup of silicate rock, which requires an energy input from the outside for the reaction to go at all. It may be slow, as in the fermentation of sugar to produce alcohol in wine. It may be fast, as in the explosive oxidation of fuel in a diesel engine.

A container of a hydrogen-oxygen mixture does not spontaneously explode. Although a great deal of energy is released when two H_2 molecules combine with one O_2 molecule, this reaction does not "go" unless the H_2 and O_2 molecules can first be dissociated into ions (two H^+ and two O^-). This dissociation consumes energy but can be initiated by a spark, and the resulting energy release then heats additional H_2 and O_2 sufficiently to dissociate them and by the chain reaction to produce an explosion. The rate-limiting step is the dissociation.

It took extensive science to determine the rate-limiting step in the production of photochemical smog in the Los Angeles basin. It was known that at least nitrogen, oxygen, water vapor, ultraviolet sunlight, ozone, and hydrocarbons were involved, but the chain of reactions was complicated and possibly other elements like smoke particles participated. It turned out that the rate-limiting step, which thereby determined how much smog would occur, had a velocity that was proportional to the concentration in the air of hydrocarbons (mostly from industrial processes and unburned hydrocarbons in automobile exhaust). Finding this rate-limiting step was clearly the key to understanding and then to reducing photochemical smog.

Identifying the rate-limiting step in some activity of a corporation or an institution can be even more difficult than in a complicated chain of chemical reactions. It can be as simple as two individuals who do not talk to one another, or as complicated as a failure to put in place proper incentives so that the feedback from a computer-dominated communications channel is used by the manager of a manufacturing process.

Closely related to operations analysis is *systems analysis*, a function and a phrase that also enjoyed a big boost from military operations. Systems, such as a worldwide accounting system in a corporation; the worldwide network of parts and people to maintain Boeing 747 aircraft; or the NATO system of aircraft, airfields, and supporting logistics, are susceptible to analysis as a whole, including but not limited to the operation of each component. In the U.S. Defense Department, it is now popular to evaluate the five-year cost of a system. There is nothing magic about "five," but including the aftermanufacture costs of training, maintenance, replacement parts, and losses by accident along with development and manufacturing costs clearly permits appropriate analysis of the whole system, its cost, and its effectiveness. Only on the basis of an analysis of the total costs over a period of years can one system compete fairly with a rival system to serve the same function.

The scientist soon learns that the Achilles heel of a sophisticated experimental apparatus is the cables and connectors. This is true both in the restricted, literal sense of cable faults and connector corrosion, but even more in the figurative sense of the interoperability of the components: Does, for example, the spectrograph "talk" to the computer in a language it understands? Agreement on languages, protocols, and standards enables a system to function, whether literally for the human language of a NATO exercise or figuratively for the voltages and waveforms of scientific equipment. The toughest problems with a system are almost invariably not in the components but in the connections, the interfaces, the compatibility of components, and the integration into a harmonious, functioning whole.

The message for an organization is obvious and usually appropriate. Achieving this integration is clearly a vital concern of executives. It is necessarily a "top-down" function, it cannot be built up from the bottom.

The publisher's copyeditor is an indispensable link between the author and the printer of a book. The copyeditor straightens out ambiguities and inconsistencies, corrects grammar and spelling, and marks the manuscript for size and style of type and for all the other choices that must be made of headings, spacings, and styles.

One such copyeditor of a college textbook cleared his desk each Friday afternoon of the batch of pages he had been working on that week and express mailed them to the author, who had to check each batch of marked manuscript before it went to the printer. The batches were duly delivered to the author each Monday morning. The author, like most college textbook writers, was a college professor, who could become efficiently immersed in the book project on weekends but who had only an occasional weekday evening to work (inefficiently) on manuscript. The manuscript therefore lay idle for five days each week.

Obviously each of these cycles that took two weeks could have been done in one week, with a great saving in the publication time, if the copyeditor had shipped batches on Thursday, and even more time could have been gained if he had adopted the practice of calling the author to learn when she could work on the manuscript. Using FedEx as a substitute for thought and then blaming the author for the slowness of the production process was costing time and therefore money.

This observation could be called operations analysis, or rate-limiting-step identification, or systems analysis (the system of copyeditor with author). No mathematical statistics was involved, and you could appropriately observe that the proposed improvement was just common sense. Yes, that is true, but it was common sense driven by the introspection and attitude of operations analysis. It is this attitude that is the crucial and enduring idea that I wish to recommend to you.

The location and sizing of new production facilities calls for operations analysis of the most general and sophisticated kind. Not only must you consider the availability and cost of labor, the in-transportation of materials, the out-transportation of products, the tax and political environment, the access to markets, and many other considerations; you must also study the first derivative of each of these and even the second derivative of some. The optimization process has many dimensions. In some of them (e.g., the demography of the labor force), mathematical statistics and modeling will be required. In others (e.g., the support or lack thereof from the local congressional representative), no amount of mathematics will help. In the end, you must avoid suboptimization along any one axis, since you are seeking the grand optimum overall.

Even if you have a sophisticated grasp of management in these terms, it is still true that management is an *art* as well as a science. Forgive me if I now digress to give two examples of the art form: The first is what I call *Scheherazade management*. You will remember Scheherazade who tells the tales in the *1001 Arabian Nights*. The sultan had resolved to behead the narrator after an evening in the seraglio, but she invented tales of such interest and then interrupted each story in such a crucial place each night that the sultan had to preserve her life for another evening in order to hear the rest of the story. Then, of course, she started another story. I have known associates who played this game successfully for several cycles, never finishing a project until well enmeshed in a new one.

The second example is what I call *Don Giovanni management*. At the opening of the second act of Mozart's opera, the Don, disguised as his own servant, gives instructions to the angry mob chasing him. Briefly stated, they are: "Half of you to the right, half to the left, and the rest follow me." I have looked on in admiration as some of my acquaintances have performed this masterpiece of delegation, but I have never succeeded myself.

Perhaps, as I have suggested, this chapter is only common sense. But it may have helped you to take a grander, more overall and integrated, view of your operations and to make you think harder about what you seek to maximize.

16

The Big Picture

Scientists commonly rise above the details of a problem and see the big picture through the use of *conservation principles*. There is only a small number of these, but each has a wide region of applicability. Each is a statement that some quantity is *conserved*, that it neither increases or decreases in the course of the operations considered. Conservation principles ordinarily apply to *isolated* systems, that is, systems that can have tremendously complicated interactions among their internal parts but which do not interact appreciably with "the outside world."

Of these, the First Law of Thermodynamics, otherwise known as the Principle of the Conservation of Energy, is the most famous.

There are several ways of stating the First Law. I am content with: Heat and work are equivalent, and in a system isolated from its surroundings either can be produced only by the reduction of the internal energy of the system. "Internal energy" can be, for example, mechanical energy stored in springs or flywheels, chemical energy stored in coal or hydrogen, or nuclear energy stored in uranium nuclei. Work done or heat produced thus comes at the expense of using up the stored energy.

Work and heat can be converted into one another, subject to some restrictions that I shall outline in the next chapter. If there is no change in the internal energy, then any increase in heat (thermal energy) occurs by the expenditure of work (a change in mechanical energy), and any work occurs by the expenditure of heat. The total energy of the isolated system remains constant; it is "conserved."

Another way of stating the First Law is to say that a perpetual motion machine is impossible. Any such machine would have frictional losses (producing heat), and to be useful as a machine it would have to do work. An isolated machine cannot produce heat or work without using up a source of internal energy. The impossibility of a perpetual motion machine is an example where the use of a conservation principle makes detailed calculations unnecessary. A generation or two ago there were numerous "inventors" claiming spectacularly attractive properties for very complicated machines they had created in concept. They all violated the First Law. Analyzing in detail exactly what was going on inside such a machine was often a very tedious process, so exhausting as to discourage critical analysis. The more talented the con man, the more complicated was his mechanism. But one could cut through all the distracting complications and be confident that the machine was flawed and the claims nonsense, since it violated the First Law.

I do not know why these "inventors" are fewer or at least quieter now. Perhaps they are practicing other confidence games that are less obvious. Probably their most dangerous tactic, which procured sympathy from the scientifically illiterate, was their righteous anger if one did not agree to analyze in detail the Byzantine complexity of their schemes (in which a fault had to be hiding some place) but instead relied on the First Law to prove that their schemes were nonsense.

One of the burdens of being a university president is the obligation to respond courteously to the 0.001%± of alumni and other friends of the university who have "invented" perpetual motion machines or other wonderful machines that violate other physical laws. Although it is always easy to use conservation principles to show that the machine does not work, it is more difficult to answer convincingly the inventor's question: "Why do you place confidence in that physical law instead of seriously studying in detail how my (brilliant) machine works?"

The answer is that the evidence for the First Law is simply overwhelming. The world would be a quite different place if it did not hold. No modern experimenter conducts experiments specifically to verify it, but since we all rely heavily on it in all our work, the slightest "softness" in the law would be exposed hourly.

A great deal of attention is being applied in the 1990s to the "greenhouse effect" and the possibility of "global warming." The former is the wholly benign way in which water vapor, carbon dioxide, and other "greenhouse gases" (of which trace amounts are present in the atmosphere) allow sunlight to reach the earth's surface but limit the reradiation of heat from the earth's surface back into space. The gases play the same role that glass does in an ordinary agricultural greenhouse. Without them, the earth's surface would be too cold for life as we know it.

There is active speculation that man-made additions to the blanket of greenhouse gases (notably carbon dioxide) may result in a small, gradual warming of the globe. In analyzing this situation one relies basically on the First Law applied to the earth as a whole. The earth is not quite an isolated system, and the insolation (solar energy incident), the part of that which is reflected from clouds, sea, and earth, and the radiation from the earth and its atmosphere (a sensitive function of temperature) all must be accounted for. Then the conversion of internal energy (primarily from wood, coal, gas, and oil burned by people but also from forest fires and the radioactivity in the earth's interior) must be calculated. But all of these and many more, once they are known with sufficient accuracy, can be put into one grand equation, which is simply the First Law, and the temperature is then derived.

In addition to the complexities noted, there are many more; for example, carbon dioxide concentrations are reduced by photosynthesis processes, primarily in forests, and by incorporation into carbonate rocks and solution in lakes and oceans, and these concentrations are presumably increased by burning fossil fuels. The number and variety of the contributions and reactions, each with a different time of producing an effect and each subject to uncertainties of measurement, have been responsible for the considerable controversy over the (preliminary) estimates being made. But there is no controversy over the First Law, which is solid, permanent, and respected.

The thought processes of dealing with an isolated system and using a conservation principle can be employed to visualize business accounting. You can draw an imaginary balloon around your company. Investment cash and revenue from sales flow into the balloon, and salaries, cost of goods, taxes, interest on debt, and dividends flow out. Cash used inside the balloon for creating facilities or intersegment transfers (the "eliminations" on corporate income statements, sales from one division to another) are like the storing or using up of internal energy, they do not flow in or out across the balloon wall. The net flow across that wall is a strong measure (but, of course, far from the only measure) of the health of your enterprise.

These same thought processes can be used to visualize the economic behavior of a country. You can draw an imaginary balloon around the United States and examine the flows of goods, services, and money into and out of the country. If the internal supply of capital (like internal energy) is not to be used up, these flows through the balloon must balance. Many prominent politicians have proposed that we need not worry about the loss of manufacturing jobs to Germany or Japan, that the service industries (dry-cleaning, banking, and the like) would provide the jobs. To the extent that some services, such as some banking, are paid for by other countries and therefore constitute a net outward flow through our balloon, these activities would help; that effect is small and is not what they have in mind. But anything like the insurance that we sell to each other inside the balloon does not help. The net inflows of oil and television sets must ultimately be balanced by sufficient out-flows, and that means selling natural resources, agricultural prod-ucts, and manufactured goods. For a while, the inflow of in-vestment capital (buying factories, real estate, and some of the national debt) can balance the books, but this is obviously only temporary.

In mechanics the *principles of conservation of momentum* are highly useful. The *linear momentum* of an object of mass *m* is the product of *m* and its velocity *v*. For a collection of masses, the linear momentum is the sum of the products of each mass and its velocity. If this collection is not acted upon by any exter-nal forces, the linear momentum of the ensemble is conserved, and this is the Law of Conservation of Linear Momentum. This law can easily be proved from Newton's Laws of Motion and has been verified in many ways. The individual masses of an isolated system can interact with great elaboration and compli-cations, with as intricate a combination of springs, gears, and sources of energy as you can imagine, but the overall momen-tum will remain constant. As in the application of the First Law, the conservation of linear momentum can be used to cut through the complexities and to analyze the motion as a whole.

If a crowd of people is advancing upon a barricade the crowd as a whole has momentum. Disagreements and even fights can break out among the participants, but as long as the crowd holds together, its momentum is conserved. Similarly, your employ-ees may (figuratively, of course) be advancing on you with some proposition or complaint. You may take some comfort from internal squabbles among them, but even if they break up into independent groups, the total momentum prevails. It is this lin-ear momentum that is the basis for the common use of "mo-mentum" when one speaks of the momentum of a sales cam-paign or of a new product, a nearly universal application of a scientific metaphor.

There is a similar Law of Conservation of Angular Momen-tum. It is similar to the Principle of Conservation of Energy in the way it is applied. But it is not so far-reaching into everything we do, and you may wish to skip the rest of this chapter.

Before introducing it, I need to introduce not only angular momentum, but also two other concepts, *angular velocity* and *moment of inertia*. Consider a particle of mass m moving with velocity v and located at a distance r from some axis (Fig. 16-1).

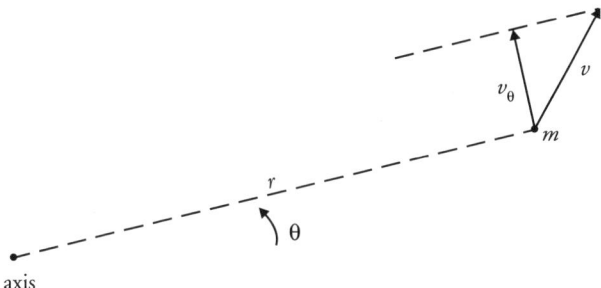

Figure 16-1. The angular velocity about the axis is v_θ/r.

I define v_θ as the component of v perpendicular to r. Otherwise stated, v_θ is the projection of v on the arc of the circle with its center at the axis. Then the angular momentum about that particular axis is defined as $L = mrv_\theta$.

If we recall the definition of the radian measure of an angle (the arc length divided by the radius), the rate of change of arc length is just v_θ. Dividing by the radius gives us the rate of change of angle, which we call the *angular velocity* ω, and then $\omega = v_\theta/r$. Thus

$$L = mrv_\theta = mr^2\omega$$

If we have a collection of masses moving with the same angular velocity (as in an automobile wheel), we can sum the contributions of each individual mass to get the total angular momentum

$$L = (m_1r_1^2 + m_2r_2^2 + m_3r_3^2 + \ldots)\,\omega$$

The sum in the parenthesis is a property of the masses and their distribution, and we call it the *moment of inertia I* of the collection of masses. The moment of inertia (again think of the automobile wheel) obviously weights most heavily those masses far from the axis (larger r's); it is their motion that contributes most to the angular momentum. You can understand the "inertia" in this name if you have ever tried to rotate an automobile wheel when the auto was raised on a jack. You can appreciate the "moment" part by understanding that the farther the mass is from the axis (larger moment), the more difficult it is to accelerate the angular motion.

Now the Law of Conservation of Angular Momentum states that if there are no forces in the θ direction (at right angles to r, such forces when multiplied by r we call "*torques*"), then $L = I\omega$ is constant. Again, this can be readily proved from Newton's Laws of Motion. You have doubtless observed one of the most dramatic demonstrations of this law in the behavior of a spinning figure skater. When spinning with only the toe of one skate on the ice (Fig. 16-2), there can be no forces in the θ direction, and thus angular momentum is conserved. As the skater changes her moment of inertia by extending or retracting her arms and one leg, her angular velocity decreases or increases to keep the angular momentum $I\omega$ constant.

(a)   (b)

Figure 16-2. Conservation of angular momentum. From Cutnell and Johnson, *Physics,* © 1989 John Wiley & Sons, Inc. Reprinted by permission of John Wiley and Sons, Inc.

The point I wish to stress here is that we need not examine the motion in detail and the complicated forces of the skater on her arms and legs in order to predict the basic phenomenon, the rotational speedup as the skater's arms and one leg are brought closer to the spin axis: We can get this result directly from the conservation principle.

A good illustration of the power of reasoning based on a conservation principle is the analysis of the winds associated with a low- or high-pressure disturbance in the earth's atmosphere. This example also illustrates how a scientist makes good use of the consideration of simple special cases of a problem to infer the result for the more general situation without having to calculate it in all its complexity (you will recall our earlier discussion of the generalization process).

Let us look down upon the earth from a position high above the North Pole (Fig. 16-3). Let us assume that at this moment there is a low atmospheric pressure disturbance centered at the pole and that the atmosphere is otherwise at rest (with respect to the earth, of course). Masses of air will accelerate from the higher to the lower pressure region, which means they will start

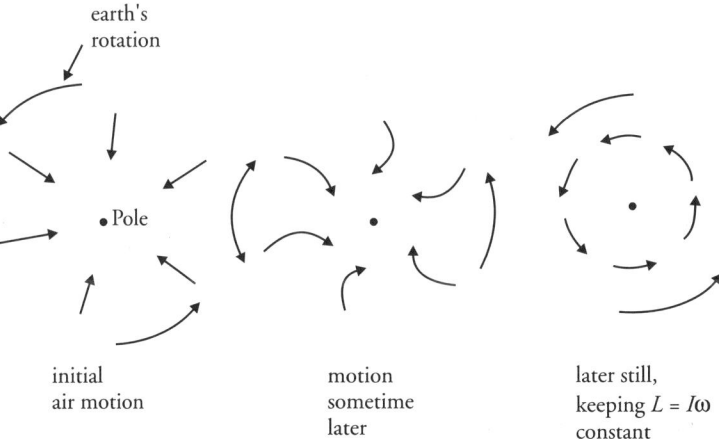

earth's rotation

• Pole

initial
air motion

motion
sometime
later

later still,
keeping $L = I\omega$
constant

Figure 16-3. Counterclockwise circulation about an atmospheric low at the North Pole.

out toward the pole (the drawing at the left in the illustration). But from our Olympian position above it all, we know that this mass of air, at rest with respect to the earth, was already rotating with the angular velocity of the earth. You can easily verify that I have the correct sense for this velocity by noting that the sun rises in the east; rotate the circle of the earth a little in the sense of the arrow and note how the (fixed) sun appears to rise above the eastern horizon.

Initially there are no forces in the θ direction, and so angular momentum is conserved. As the air masses move in toward the "low," the moment of inertia of the air masses decreases (the average r^2 decreases), and to keep $L = I\omega$ constant the angular velocity of the air masses must increase. Therefore the air will appear to move relative to the earth in a counterclockwise circulation (the middle drawing in Fig. 16-3).

As the motion continues, forces develop in the θ direction, but these are just the drag of trees (well, where we are, ice) on the wind, which reduces the effect quantitatively without altering its basic character. We have thus established that about a low at the North Pole there develops a counterclockwise circulation in which the winds do not just approach the low's center but move almost at right angles to that direction and in the counterclockwise sense.

We really do not care much about the North Pole. But this big cyclonic circulation we have discovered cannot change much as we go from 90° to 80° or lower north latitudes. The effective ω of the earth becomes somewhat less and the counterclockwise circulation becomes somewhat less strong, but it does not go away. Thus at temperate latitudes (~30° to ~60°), we can expect a counterclockwise set of winds around a low-pressure area.

If we imagine going to even lower latitudes, however, by the time we get to the equator, the effect has disappeared since the air masses no longer have an initial angular velocity (Fig. 16-4). And it is easy to see by making another sketch like my first (keep the sun rising in the east!) that in the *southern* hemisphere the circulation about a low will be *clockwise*. Likewise we could

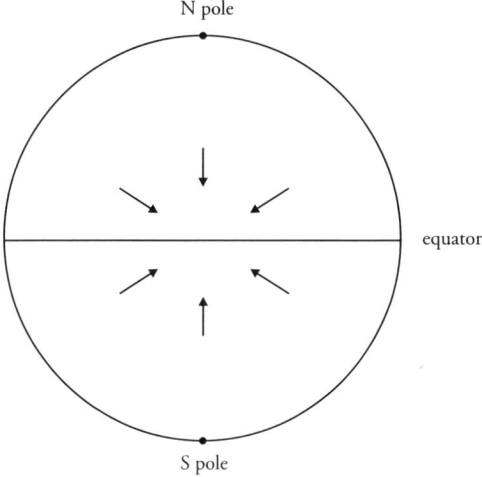

Figure 16-4. The effect goes away at the equator.

see that a circulation about a "high" would be clockwise in the northern hemisphere and counterclockwise in the southern.

Thus by examining a few special cases and using the conservation principle, we have developed an important meteorological law, called Buys-Ballot's Law (you must use the French pronunciation, else it sounds like a cynical comment on the electoral process!).

You probably have already heard of this law, sometimes with wholly unproved applications. I should therefore caution against too quick extension of this law to vortices in drains, sinks, or bathtubs: The radii in these cases are smaller by a factor of about 10^8 than in the atmospheric circulation, with consequently much smaller moments of inertia. In the tub, secondary effects like waves at the time the plug is pulled and deviations from perfect circular symmetry of the opening dominate the motion and determine the sense of the vortex.

You can appreciate from this chapter that when a system can be, to a good approximation at least, isolated from its surroundings, conservation principles apply. Analyzing the behavior of the system on the basis of conservation of energy, momentum, or angular momentum often is far easier and more illuminating than detailed analysis of its internal workings. This mode of thought can be applied to corporations or organizations or even countries.

You have also seen (if you managed to penetrate the second half of this chapter) an example of how the scientist can develop a general result from the analysis of special cases, frequently (as here) without the use of mathematics. You have probably used this process of generalizing in applying the results of test marketing in limited areas to predict the results of marketing on a worldwide scale.

17

Sloppiness Is on the Increase

Perfect order is never achieved. In popular language there are many words for kinds of disorder, words such as "sloppiness," "untidiness," "disarray," and "vagueness." An executive must actively limit disorder in the face of the almost universal growth of it.

The scientist's measure of disorder is called *entropy*. The entropy of a highly ordered system such as a crystal of ice increases as the ice melts into a disordered liquid. If I mix the contents of a bottle of pure oxygen with one of pure nitrogen, entropy increases.

In an electricity-generating station, energy that was originally concentrated in coal and then in a relatively small amount of high-temperature steam is used to drive a steam-turbine electrical-alternator combination, giving off waste heat to a large reservoir of cooling water. Although engineers have designed the plant to have the highest possible efficiency, the limitations of cost and availability of materials have constrained their design. Much of the energy becomes available in electricity, but most of it is necessarily lost into a form (warmed cooling water) that is useless for delivering useful work. Entropy has increased.

Entropy is closely related to the Second Law of Thermodynamics. As with the First Law, many alternative statements are possible. One is that heat in any system always flows from the warmer to the cooler regions; the warmer regions cool and the cooler regions warm. Another statement is that one cannot get work out of a system by cooling the cooler region, even though that would not violate the First Law. Left to itself, a system would finally have a uniform temperature and there would be no possibility of extracting work from heat.

Another, equivalent statement is: In the course of any process, the entropy of the universe either stays the same (if the process is a precisely reversible one) or increases (if any element of irreversibility is present). An example of an *almost* reversible process is the pendulum: I start with the mass pulled out and up from its lowest position. From the time of release, the velocity (and hence the kinetic energy) of the mass increases toward the low point and then decreases to zero when the mass (almost) reaches the height from which it started; this motion repeats over and over again.

The process is not quite reversible because some of the energy is dissipated in the air resistance, in heat generated by friction at the point of support, and in other second-order processes. Entropy gradually increases because of the (very slight) irreversibility, and the mass eventually settles down to the lowest point of its swing. The total entropy of the universe has increased because energy has been taken from focused, concentrated form (my lifting the mass) and dissipated into a slight warming of great quantities of air.

One of the more unfortunate characteristics of modern civilization in a "developed" country is the explosion of entropy by the way we treat our waste. In a home, old newspapers are in a pile in the living room, recyclable paper is in a wastebasket in the study, separate piles of recyclable cans, bottles, and perhaps plastics are in the garage, and biologically degradable food products are in the kitchen garbage can. So far, so good; low entropy. But then all of this is tossed into a single garbage can, indiscriminately mixed with others' wastes, and added to an already burgeoning landfill. Entropy has increased enormously and at least to some extent unnecessarily, and the waste problem is thereby made vastly more intractable than if the sorting could have been preserved. Recycling helps to some extent but is costly because it is labor-intensive and because food waste is no longer fed to pigs.

There is an old proverb, which I translate from the Catalan into contemporary English: "If you want a husband who's slim and clean, marry a man who's slim and clean. If you want a husband who's fat and dirty, marry a man who's slim and clean; he'll get fat and dirty." Entropy is at work here; sloppiness steadily and inexorably increases.

A revealing illustration of entropy is the difference between parking and unparking an automobile. Suppose (Fig. 17-1) I parallel park between two other cars that leave me a space only a little longer than the length of my car and that remain fixed during my operations. I will ordinarily move my car forward and backward several times, with hopeful spinning of the steering wheel between and during motions, and I will finally attain the required position close to the curb.

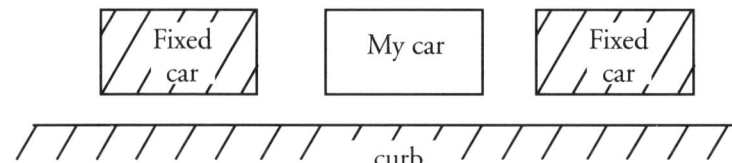

Figure 17-1. Parallel parking.

Now let us assume that the same two cars are there and in the same positions when I wish to leave. Unlike the parking process, when I leave I typically back up *once* with the steering wheel turned hard right, then go forward *once* with the wheel hard left, and in two operations I have left the parking spot! Why the asymmetry? Why is it so much harder to park than to unpark? If I had on entering simply made the exact reverse of the path I used on exiting, I could have entered with the simplicity and ease that I enjoyed when I exited.

The key to this paradox is that on entering I am required to get into a single, narrowly confined state, a position of low entropy (highly ordered). On exiting, I can go into any of a large number of states (highly disordered), with my car 1 foot, 2 feet, or greater distances from the other parked cars and at an angle of 10°, 20°, or greater angles with respect to the curb. All of these final states are successful, and the large number of possibilities means large entropy. It is like shooting an arrow to *some* place on the target compared to hitting the bull's-eye. Going from low entropy to high entropy is easy; Nature does it by herself with no human intervention or supply of energy or information. Going the other way is hard.

In the 1960s and 1970s, undergraduates at many colleges and universities enjoyed taunting their administrations with co-ed dormitory proposals. A frequent arrangement of men's (M) and women's (W) rooms had been as shown in Fig. 17-2 in

M	M	M	M	M	M	W	W	W	W	W	W

CORRIDOR

M	M	M	M	M	M	W	W	W	W	W	W

Figure 17-2. An ordered array.

response to earlier pleas for co-ed dormitories. At least this is the way the rooms were distributed at the *start* of the school year; entropy only gets larger, and at the end of the school year, the arrangement was probably much less ordered! The undergraduate claim was that this was an "unnatural" (excessively low entropy) arrangement and that the "natural" arrangement was as in Fig. 17-3. But this is just as "unnatural" and ordered; there really is only one choice, only two possibilities: Either there is an M at the left end or a W. After that choice is made, everything is fixed, just as in the earlier corridor.

Entropy in your company or in any other organization is the measure of its disorder. In many organizations a certain amount

| M | W | M | W | M | W | M | W | M | W | M | W |

| M | W | M | W | M | W | M | W | M | W | M | W |

Figure 17-3. An array that is just as ordered as Figure 17-2.

of entropy is desirable, since flexibility and the exercise of imagination can be constrained by an excessively ordered operation. But management is invariably necessary in order to concentrate the effort of the organization and to focus its attention on specific goals. This activity of management is often called contributing "negative entropy." Just how much negative entropy to apply and how to apply it are sensitive executive decisions. Leadership always means contributing negative entropy. Entropy increases when authority and responsibility are diffused by defective organization structures or by management by committee.

The Second Law gives a direction for "time's arrow." This is most obvious if we use the statement of increasing entropy as time moves onward. If I am given a group of photographs of a child's sandbox taken at different times, I have no difficulty in arranging them in the order of earlier time to later time, provided that there has been no adult intervention. Systems always degrade like this unless negative entropy has been introduced, and time moves in only one direction, the direction of increasing entropy. In our region of the universe and over spans like our lifetimes, the increase in entropy tells us how time's arrow is pointing, and there can be no nonsense about "time machines" or reversing the movement of time. (I concede that there are some deeper, philosophical problems if one's attention goes back in time to the "big bang" or goes off in space to the limits of the universe.)

There is an important connection between entropy and information. A hint of this was in the garbage problem: Information was needed to sort the various elements. If I tell you that a system is in (say) the 175th state when it has 1000 possibilities, I am giving you more information than if I tell you it is in the second state when there are only 2 possibilities. Entropy is a measure of the number of choices, of the number of different ways of getting to the observed state.

The capacity of communication channels, the rate of transmission of information, can be analyzed in terms of entropy. There are fascinating connections among entropy, the Second Law, temperature, and noise in communication channels. In the end, the fundamental, irreducible noise from the fluctuations required by entropy and the Second Law sets the limit to the rate at which information can be transferred.

The connection between entropy and choices raises one of the most (some would say *the* most) interesting problems of all time, the "mind-brain problem." A great deal is known about the physiology of the brain, its receipt of information from sensors, its storage and processing of information, and its transmission of signals to produce motor activity. Psychologists know a great deal about behavior and what choices different people make, which are surely the results of actions by the mind. But virtually nothing is known about the interaction between mind and brain and how and to what extent the mind controls the brain. Never mind for the moment big questions like value judgments, moral choices, and love. Just consider *attention*: What tells the brain which, of innumerable choices, to pay attention to next? Should I write another sentence or look out the window? What is the meaning of "I" anyway?

There has been tremendous progress in understanding the brain, but the mind-brain problem is still in about the form it was left by Descartes and his famous "I think, therefore I am." If you have trouble with an arrogant scientist who would like you to believe that scientists well understand the universe and have things tidied up nicely, I suggest you ask him what progress he and his colleagues have made on the mind-brain problem.

We considered managing energy in Chapter 16 and managing entropy in this chapter. It is not obvious how much disorder is optimal in an organization or in one's personal life, but introspection about it is always helpful. Furthermore, executive intervention is likely to be required to reduce disorder, since the natural tendency is for it to increase without bound.

18

Matchmaking

In this chapter I shall describe the concept of *impedance matching*, which comes to us from electrical engineering, and give some applications. This concept guides the transfer of energy, information, or anything else from one person to another or from one organization to another.

The concept of impedance matching can be introduced crudely by the following examples: Suppose a 250-pound professional football player is pushing a children's toy. He has no difficulty moving it, but he is hardly "in gear" with his task in a way that he can transfer appreciable energy to it. Suppose a small child is pushing on a stalled or slowly moving automobile. His tiny force is doing almost no work; it could be doing more if it were applied to accelerating a faster-moving toy.

These examples give a rough introduction to the basic idea, but I now wish to explain the concept more carefully and to tell how the name arises. To do this, I must start with some explanations of electrical circuits.

Consider a flashlight battery with a nominal output voltage of 1.50 volts. If I connect its terminals to a 1000-ohm resistor, a current of $1.50 \div 1000 = 0.0015$ ampere flows (Fig. 18-1). (This is by the application of one of the most familiar formulas in physics or everyday life, Ohm's Law: Current equals voltage divided by resistance.)

Now suppose I short-circuit the battery, that is, I connect it

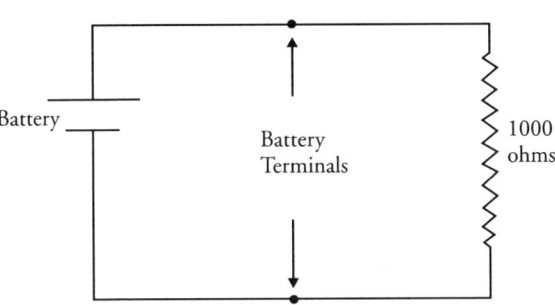

Figure 18-1. A battery connected to a resistance, the "load".

to a zero-resistance load. If the voltage at the battery terminals remained at 1.50 volts, *infinite* current would flow (1.50 ÷ 0 = ∞). It does not, because the battery source of voltage has also an *internal resistance*; the current is limited to a finite amount by the finite amount of the chemicals reacting within the battery cell. The larger the battery, the lower the internal resistance. You can momentarily short-circuit a "D" cell and draw 10 amperes but only 1 ampere from a (much smaller) short-circuited "AA" cell. (You will quickly ruin the battery unless you immediately disconnect it.) The full 1.50 volts no longer appear at the terminals of the battery. The 1.50 volts are still characteristic of the chemical reaction, but some of the voltage is lost before it gets to the terminals of the battery.

Thus we really should draw the circuit as in Fig. 18-2: The

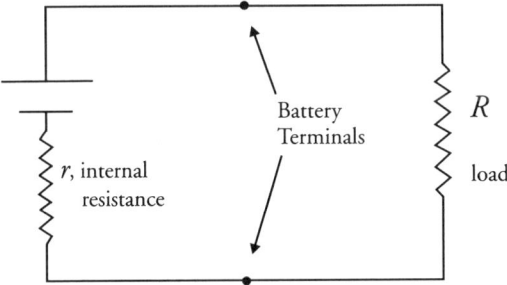

Figure 18-2. The actual circuit of Figure 18-1, including the limitations of the battery.

current that flows is $I = 1.5 \div (r + R)$. Now I ask: What should R be if I wish to transfer the *maximum power* to it?

The power P delivered to the load R is I^2R, the product of the voltage IR that appears at the terminals of the resistor and the current I that flows through it. Thus $P = I^2R = 1.5^2R/(r+R)^2$. Since r is fixed by the size of the battery, the maximum of P comes at the same value of R as the maximum of $P/1.5^2 = Rr/(r+R)^2 = R/r \div (1 + R/r)^2$. I plot this as a function of R/r in Fig. 18-3. The maximum occurs when $R/r = 1$, that is when the external resistance R equals the internal resistance r. By *matching* the resistances we have maximized the power transferred to the load.

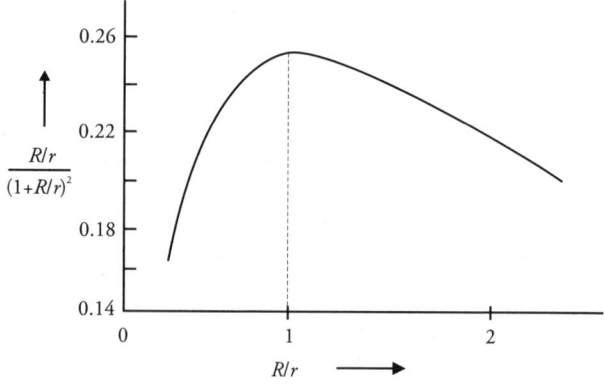

Figure 18-3. The maximum power into the load is obtained by making $R = r$.

Several footnotes: First, our maximization process would have been much easier if we had identified the maximum as the point at which the first derivative was zero and calculated that derivative. I have promised to be a good boy and not to try to teach calculus in this book, but I give the result. The first derivative dP/dR is calculated easily to be $1.5^2 (r^2 - R^2) \div (r + R)^4$, and setting that equal to zero gives $r = R$ as the location of the maximum.

Second, we have established a general result. Using the example of 1.5 volts has not in any way influenced the conclusion that resistances must be matched for maximum power transfer.

Third, I must remind you that we are speaking of maximum power transfer. This is not the way you ordinarily use a battery, since if you let $r = R$ and since the same current flows in each resistance, $I^2r = I^2R$, you are dissipating as much power inside the battery (generating heat) as you are in the load (lighting a flashlight bulb, for example). Ordinarily you will make $R \gg r$ and get much higher efficiency and longer battery life.

Fourth, I have for simplicity been speaking of *direct current* circuits. The same basic approach can be used and the same result occurs when I consider *alternating current* circuits, such as the 120 volts A.C. from the wall outlet or the radio-frequency alternating current in a radio or TV antenna. In the A.C. case, we speak of *impedance* instead of resistance; impedance includes the effects of inductance and capacitance, which we need not explore, but the heart of impedance is still the resistance and the current is still the voltage divided by the impedance.

Impedance matching is vital in drawing signals out of the sky for television or communications. We do not care about efficiency; all we wish is to obtain the maximum power into our set in order to dominate over the fundamental noise, the fluctuations in voltage required by the Second Law of Thermodynamics. In the language we used earlier, we seek the maximum signal-to-noise ratio. A TV antenna, made up of 3/8-inch diameter aluminum tubes, looks like a dead short (which it virtually is for direct current), but we must remember that the impedance for the high-frequency alternating current of the TV signal is more complicated. The shape and size of the antenna establish its impedance at 300 ohms. We therefore construct the TV receiver so that the impedance we see "looking into" it at the antenna terminals is 300 ohms.

You can now appreciate the examples in the opening paragraph of impedance mismatch and accordingly low power transfer. The low-impedance athlete cannot do much work on the high impedance toy, and the high impedance child is likewise ineffective in pushing the low impedance automobile.

In electrical A.C. circuits, if we wish to connect a source of power with a high voltage and high internal impedance to a load with low impedance, we use a *transformer*. The transformer does not add power (it actually subtracts a little), but it makes it possible to deliver more power to the load by providing a better impedance match.

We are accustomed to the use of transformers in electrical power distribution networks. For efficiency, to keep the lost power I^2R low in the transmission lines, long-distance transmission of power is carried out at high voltage and low current, that is, at high impedance. Near your home a transformer reduces the voltage and makes high current possible, delivering power to your home at low impedance (of order of magnitude 10 to 100 ohms).

In mechanical power trains, gears play the role of the transformer. The analog of current is torque and that of voltage is velocity. An automobile power train starts with a high engine (angular) velocity at a relatively low torque, and the transmission gears transform this power to a high torque at a relatively low velocity; we are in "low gear" at the start, a high gear ratio. This is the analog of a high electrical impedance being connected through an impedance matching transformer to a low impedance load. As the car's velocity increases, I shift gears (use a smaller transformer ratio) to keep matching the engine to the load.

The metaphoric use of impedance matching is widespread in a broad array of applications. For example, I say that a Nobel-Prize-winning microbiologist is not a good impedance match to digging a vegetable garden. He may do it very well and even enjoy it, but he is not "in gear" with this task in a way that fully utilizes his competence and capabilities. And I have direct and extensively documented evidence that I am a ridiculously poor impedance match to the Medicare system, but noise and limited bandwidth are other complicating features of that interaction.

Organizations can be good or poor impedance matches to

one another. A particular management style can be a good or poor impedance match to an organization's troops or problems.

An individual who serves the role of matching transformer can be extremely valuable. He or she serves to maximize the transfer between people of very different backgrounds, training, goals, or interests. Many executives have special, often roving, assistants to do just this. A copyeditor is the publisher's way of providing an impedance match between author and printer.

One of the serious problems with the late Vietnam adventure was the great impedance *mis*match between the sophisticated U.S. force and the primitive wet jungle conditions. The United States had a relatively small number of highly capable people and weapons useful mostly in the daylight, but the war required a large number of less capable people and weapons operating mostly at night. The B-52s bombed Hanoi not because it was useful but because the aircraft were capable of doing it and the crews had been trained with such a mission in view. (Mismatch was not, of course, the only problem: perhaps the largest problem was that the United States was nominally supporting the central government, but beyond a few miles from Saigon no one had ever heard of a central government.) In contrast, in the Gulf war with Iraq, the open country, better weather, and more "high-value" targets were a better impedance match to the American "high-tech" weapons.

The mismatch that one occasionally witnesses between a speaker and an audience can be a painful experience. It is not uncommon between a professor and students. I remember one freshman at Cornell who by a scheduling mistake was attending a course in Advanced Quantum Mechanics instead of introductory physics. When we asked him whether he was not aware that something was wrong, he replied that he was just as snowed in all his other courses. Is my writing a mismatch to my audience? I worry, but I go on writing.

I have already written in Chapter 8 about the clash between

producer's language and consumer's language, another example of impedance mismatch.

The analogy should not be stretched too far, but I nevertheless remind you that under matched conditions the power dissipated in the source equals that dissipated in the "load" (application). The speaker and the listener must both work hard for satisfactory articulation, perhaps equally hard. The student who upon hearing a great lecturer thinks that the new knowledge and wisdom are merely dropped on him with no effort required on his part is in for a rude awakening at examination time.

I take advantage of the fact that I have already strayed into education to describe here an impedance mismatch of a somewhat different kind, but one that dramatically portrays the fundamental difference between learning and other forms of human activity. A physics graduate student was given the lowest grade that had ever been given on the examination to sort students into those who should go on for a doctorate and those who should not. The student had been forced by family tragedies to work in and eventually to run the substantial family business for twenty years after his bachelor's degree. After making a great success of that, he entered graduate school to do what he had always wanted to do. The long gap put him at a great disadvantage in his graduate courses. His advisor was nonplused at the examination result and asked the student what *he* thought he should do. The answer: "I don't know. Every time I've been in this situation before I've put three more men on the job."

To return to impedance matching, I remind you that there is much concern about and attention to *technology transfer*. This is the process of applying the new technology developed in one place or organization to produce a new process or product in another place or organization. It is much more difficult than most people realize and often is more difficult and takes longer than the development of the technology. The impedance mismatch between developers and appliers is most of the problem, but geographical separation can also participate. Again, talented people serving as matching transformers are vital. Parenthetically, I hope that this book may help both the technical and the nontechnical people in this process understand one another and work effectively together.

Since we have gone to the trouble of studying electrical circuits, I should like to harvest one other concept from them, even though it is not part of impedance matching. I show in Fig. 18-4 an example of an electrical circuit in which the two elements are in *series*. (We already saw such a circuit at the beginning of this chapter.) If either one of the resistors has a very large resistance, a very small current flows.

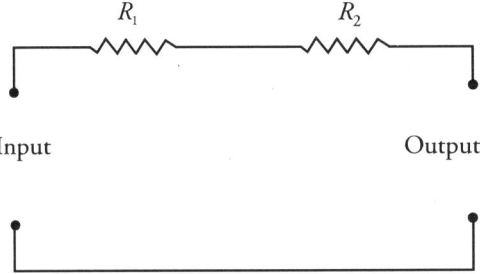

Figure 18-4. A series circuit.

I contrast this circuit with the circuit in Fig. 18-5, in which the elements are in *parallel*. Now even if one of the resistances is large, the current flows unimpeded through the other element.

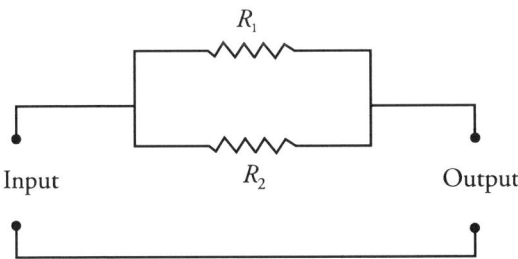

Figure 18-5. A parallel circuit.

The metaphoric application is almost obvious. A single individual who is in series with a project or program and is unfriendly to it or incompetent can thoroughly impede any motion. If that individual can be shunted aside into a parallel position, he can help if he chooses to do so but cannot stop the flow of activity in parallel channels.

This brings up in my mind the subject of *committees*. There are many circumstances in an organization where appointing a committee can be useful. It can bring the wisdom and experience of diverse individuals to bear on a complicated problem. Even a "Noah's Ark" committee, composed of one from each "politically correct" population or one from each department or product line of a corporation, can help to get data and warn of unintended consequences of projected actions. But the committee should be in *parallel*, not in series, with the activities of the person responsible for action. (Except, of course, in the case where the purpose of appointing the committee, all too common in the federal government, is to delay action.)

As you see, the difference between series and parallel is more than just the difference between sequential and simultaneous. The series-parallel contrast is in the structure of organizations and projects.

You will have seen in this chapter the importance of impedance matching in the broadest sense: speaker to audience, author to reader, executive to associates, executive to management tasks, and so on. Nothing happens without some matching, and imaginative attention to impedance matching can substantially enhance the effectiveness of an individual or an organization.

19

It's a Quantitative World

No new metaphor is introduced in this chapter, only some comments on the different meanings attached to the adjectives *qualitative* and *quantitative* and some other, concluding comments. I was very slow in learning that scientists and nonscientists attribute quite different meanings to these adjectives. It will help you in your interactions with the world of science and engineering to understand the difference.

First, the nonscientist's usage, and I admit in advance that I exaggerate considerably: "Qualitative" describes the thought and feeling of humanists and other "real people." Values, esthetics, emotions, morals, beauty, and other "good" elements are front and center in any consideration. "Quantitative" describes the nonfeeling thought of the number-happy scientist. For the nonscientist, quantitative is a distinctly pejorative adjective.

In contrast, the scientist's usage does not apply to people at all but to the stages of learning about nature. The scientist claims that he or she possesses all of those same good elements, too. The scientist's usage is an extension of the familiar division of analytical chemistry into qualitative and quantitative analysis. In the former, I determine by appropriate tests what the constituents of some unknown mixture are. In the latter, I do all of that *plus* I determine *how much* of each is present. For the scientist, achieving a quantitative understanding is the next step in knowledge and, far from a pejorative connotation, is highly respected.

The two words thus separate the population into two camps, C. P. Snow's "two cultures." From here on, I will keep to the scientist's usage. A qualitative understanding of some physical process reveals what the important elements are. In the example of the pendulum, it tells us that the length of the wire and the acceleration of gravity are the key elements, and that the longer the wire, the longer the period of oscillation. It tells us also that the mass of the bob is unimportant (within limits).

The quantitative understanding absorbs and utilizes the qualitative understanding and goes one giant step farther to reveal that the period of oscillation is $2\pi(l/g)^{1/2}$ (where l is the length of the wire and g is the acceleration of gravity). The quantitative law sacrifices none of the qualitative understanding and only *adds* to it.

Our discussion of Buys-Ballot's Law was qualitative. It produced the sense of the circulation and the directions of the winds about a low or a high. The next step in understanding would be to add to this and to relate the *magnitude* of the wind velocity to the rate of change of barometric pressure. When this is done (with mathematics and fluid dynamics that we have not gone into), we have a quantitative understanding in what is called the "geostrophic wind equation."

I bring this whole question up because nonscientists sometimes use *their* definition of "quantitative" to imply insensitivity or even callousness on the part of scientists who are undertaking quantitative thought. The careful choice of usage becomes especially important in controversies over safety, health, or environmental impact.

There is no difference between the qualitative understandings held by nonscientists and scientists about the effects of microwaves on people. We all agree that people are made of meat and that sufficiently strong microwaves will cook hamburger. We all agree that human health and wellbeing deserve protection. The scientist goes beyond to evaluate just how strong the microwave energy leakage from a microwave oven must be to constitute an appreciable threat to human health. He or she then uses operations analysis to evaluate the value to the cook of an easily opened door and the trade-offs among convenience, cost, reliability, and risk. Achieving this quantitative understanding does not imply any lack of concern by the scientist for the quality of life or any lack of appreciation of the higher values the humanist associates with his use of "qualitative."

It is a quantitative world, and if one does not recognize that, he or she can produce great difficulties and complications. The Delaney clause of the 1958 amendment to the Food, Drug and Cosmetic Act states that no additive may exist in any food if it produces cancer when fed to a human or laboratory animal in any concentration. In the nearly forty years since that legislation was passed, chemical analytical techniques have developed spectacularly, and now almost every chemical species can be found in almost every food. The Delaney clause, which is still on the books, therefore gives no guidance at all as to how large a concentration of a chemical can be tolerated in food if a large dose gives cancer to an occasional laboratory rat.

Beauty, as is well known, is in the eye of the beholder. I enjoy beauty in a vast variety of natural and artificial settings. Of the latter, music is certainly one of the finest. Even after a hard workday I can enjoy serious music, and I have made considerable effort to achieve the understanding that enhances enjoyment. I am sad that I have not purchased the recordings necessary to appreciate the music of the great contemporary composers, since I can hardly achieve an impedance match in the single public performance that I hear. I am even somewhat embarrassed that I have put my priorities for the use of time elsewhere. But then I cannot claim to have harvested for my enjoyment even all that is in J. S. Bach, including mathematics. More effort would have produced more impact. I believe that most scientists are even more committed than I am to enjoying the enhancement of the quality of life that comes from music.

The late Richard Feynman tells in *What Do You Care What Other People Think?* (W. W. Norton, New York, 1988, p. 11) of a conversation with an artist friend who puts down "the scientist" as one who cannot see the beauty of a flower, who takes "it all apart and it becomes dull." Feynman claims he, too, sees the beauty, but in addition he appreciates the mystery of flowers' evolving colors to attract insects, which therefore must see in colors. This prompts him to ask: Could insects have an "aesthetic sense?" He concludes: "There are all kinds of interesting questions that come from a knowledge of science, which only adds to the excitement and mystery and awe of a flower. It only adds. I don't understand how it subtracts."

A mathematician speaks of some particularly fine mathematics as "elegant," and a theoretical physicist speaks of a great theory as "beautiful." I can hardly expect that after a hard workday you can derive pleasure from studying some remarkable scientific success expressed in partial differential equations, but I urge you to think about this intellectual activity as capable of producing beauty and deserving of *some* effort of appreciation. Do you see, as I do, some esthetic quality in Buys-Ballot's Law? I urge you *not* to believe in an esthetic pecking order separating the two cultures, with nonscientists occupying the higher ground, and I urge you *not* to denigrate everything quantitative.

If a management meeting is not going anywhere and still consuming time, others may be vaguely upset, but the quantitative scientist will likely calculate in his head the salary cost of the exercise. He or she will then go beyond that to estimate (of course only very crudely) the additional cost in diverting attention from more consequential work.

Since I am well under way in defending the scientist, I may as well continue with one more point. A popular image of the scientist portrays him or her as arrogant. I suppose this comes from the fact that some facts, relations, and laws really are *known*, and there is no point in pussyfooting around them. But I am afraid that the image is nurtured by other interactions and pronouncements when we cannot claim this justification. Nevertheless, I maintain that science is an especially *humbling* profession. Nature guards her secrets most carefully, and when we achieve an understanding of one of them, two more usually arise to humble us.

Furthermore, the heart of scientific behavior is the acknowledgment of the capacity for error: I must subject my work to intense scrutiny, to imagine all of the errors I might have made. I must tell colleagues and critics exactly what I did so that they can reproduce it, and I must obtain their objections and deal carefully with them before I claim a new result. And even then it may be wrong, and I can only hope that I will discover the error before others do.

Parenthetically, humility appears in another unlikely embodiment: college professors. After striving all semester to get across some important understanding, they give examinations and learn how miserably they have failed. I wish newspaper writers would invite the same humbling experience by asking their readers occasionally what they have learned from a published story.

Finally, a few words about a *physical theory*. The word "theory" is used and misused so casually that you may not realize how the scientist uses it. You will read about the "police theory" of a particular crime, for example, when "hypothesis" or even "guess" would be a more appropriate word. One frequently sees something like "theoretically so and so, but actually such and such" in print; you should read this as "theoretically on the basis of a bum theory."

A physical theory is a quantitative relation among the variables in a physical situation, like the pendulum we considered earlier. It is important that the scientist who creates the theory not believe it until *predictions* have been made from it and the predicted behavior compared favorably with experiment. A sterile theory, unable to make predictions, is of no value. A theory that has survived critical tests by experiments can be confidently applied in new situations. If properly applied, "theoretically" and "actually" will agree.

Even a theory that we know is wrong (i.e., does not withstand the test of experiment) can be useful, provided we emphasize at every step of the way that the theory is incorrect. Victor Weisskopf tells a story about the pre-World War II Austrian trains, which were notoriously late. He once said to the conductor: "Why do you have timetables, since the trains are always late?" The conductor replied: "We must have timetables; otherwise how could we tell how late the trains are?" Weisskopf justifies the use of a theory (even though it is known to be incorrect) as a framework that permits experimenters to compare their results by specifying in what way and by how much his or her results differ from the (incorrect) theory.

At the present state of much of the social sciences, many "theories" do not predict, they are merely *explanations* of data that have already been obtained. Such a sterile theory must be viewed with considerable skepticism. (Recall the fallacy of the stock market "predictions" mentioned in Chapter 5.) Even weaker language has crept into some of the social sciences, where it is said that a theory is an *interpretation* of observations.

I return to literature as a coda to this book, which started with a brief reference to literature in the Introduction. The understanding and enjoyment of literature has not suffered by the addition of quantitative study, as in the authentication of authorship by computer-based word counts. Of course, analysis can be carried too far, as the late Dr. Lewis Thomas has pointed out in gently deriding both his own profession (M.D.'s) and overly quantitative teachers of poetry: "All most poems really need is a little fresh air and bed rest."

Index